08 J4046178

D0491414

UNI... AT MEDWAY LIBRARY

UNIVERSITY OF GREENWICH LIBRARY
Ju6/
HON

NW DON

TUTORIAL CHEMISTRY TEXTS

3
Main Group Chemistry

WILLIAM HENDERSON

University of Waikato

RS•C

ROYAL SOCIETY OF CHEMISTRY

Cover images © Murray Robertson/visual elements 1998–99, taken from the
109 Visual Elements Periodic Table, available at www.chemsoc.org/viselements

ISBN 0-85404-617-8

A catalogue record for this book is available from the British Library

© The Royal Society of Chemistry 2000

All rights reserved
Apart from any fair dealing for the purposes of research or private study, or criticism or
review as permitted under the terms of the UK Copyright, Designs and Patents Act,
1988, this publication may not be reproduced, stored or transmitted, in any form or by
any means, without the prior permission in writing of The Royal Society of Chemistry,
or in the case of reprographic reproduction only in accordance with the terms of the
licences issued by the Copyright Licensing Agency in the UK, or in accordance with the
terms of the licences issued by the appropriate Reproduction Rights Organization out-
side the UK. Enquiries concerning reproduction outside the terms stated here should be
sent to The Royal Society of Chemistry at the address printed on this page.

Published by The Royal Society of Chemistry, Thomas Graham House, Science Park,
Milton Road, Cambridge CB4 0WF, UK
For further information see our web site at www.rsc.org

Typeset in Great Britain by Wyvern 21, Bristol
Printed and bound by Polestar Wheatons Ltd, Exeter

Preface

WITHDRAWN FROM UNIVERSITIES AT MEDWAY LIBRARY

More than in any other region of the Periodic Table, the main group (s- and p-block) elements are diverse, ranging from highly reactive non-metallic elements such as fluorine, through semi-metals (*e.g.* silicon) to the highly reactive alkali metals. The Periodic Table itself is an excellent framework for the discussion of the trends in physical and chemical properties of the main group elements and their compounds. This book has been written with the aim of summarizing some of the more important features of the vast and varied chemistry of the main group elements, emphasizing some of the more important periodic trends and principles for rationalizing the properties. Discussion of the descriptive chemistry is backed up by worked examples and problems. Space restrictions have limited discussion of several areas where main group elements are significant players, such as bio-inorganic chemistry and industrial chemistry.

I would like to thank Professor Brian Nicholson, Dr. Richard Coll and Dr. Michael Taylor for many helpful comments on various drafts of the manuscript. I am also indebted to my wife, Angela, for giving me a teacher's point of view on the material.

Further problems, with answers, are available on the RSC's Tutorial Chemistry Texts website at http://www.chemsoc.org/tct/maingrouphome.htm

Bill Henderson
Hamilton, New Zealand

TUTORIAL CHEMISTRY TEXTS

EDITOR-IN-CHIEF:

Professor E W Abel

EXECUTIVE EDITORS:

Professor A G Davies
Professor D Phillips
Professor J D Woollins

EDUCATIONAL CONSULTANT:

Mr M Berry

This series of books consists of short, single-topic or modular texts, concentrating on the fundamental areas of chemistry taught in undergraduate science courses. Each book provides a concise account of the basic principles underlying a given subject, embodying an independent-learning philosophy and including worked examples. The one topic, one book approach ensures that the series is adaptable to chemistry courses across a variety of institutions.

TITLES IN THE SERIES

Stereochemistry *D G Morris*
Reactions and Characterization of Solids
  S E Dann
Main Group Chemistry *W Henderson*
Quantum Mechanics for Chemists
  D O Hayward
Organotransition Metal Chemistry *A F Hill*
Functional Group Chemistry *J R Hanson*
Thermodynamics and Statistical Mechanics
  J D Gale and J M Seddon
d- and f-Block Chemistry *C Jones*
Mechanisms in Organic Reactions
  R A Jackson

FORTHCOMING TITLES

Molecular Interactions
Reaction Kinetics
Electrochemistry
X-ray Crystallography
Lanthanide and Actinide Elements
Maths for Chemists
Structure and Bonding
Bioinorganic Chemistry
Spectroscopy: Physical
Spectroscopy: Organic
Biology for Chemists

Further information about this series is available at www.chemsoc.org/tct

Orders and enquiries should be sent to:
Sales and Customer Care, Royal Society of Chemistry, Thomas Graham House,
Science Park, Milton Road, Cambridge CB4 0WF, UK

Tel: +44 1223 432360; Fax: +44 1223 423429; Email: sales@rsc.org

Contents

1

Some Aspects of Structure and Bonding in Main Group Chemistry

Aims

By the end of this chapter you should understand:

- The concepts of ionization energy and electronegativity in surveying the periodic trends in the properties of the oxides, chlorides and hydrides of the main group elements
- The use of valence shell electron pair repulsion theory (VSEPR) in predicting molecular shapes, and basic molecular orbital (MO) theory for describing the bonding in diatomic molecules

1.1 Introduction

As a prelude to discussion of the chemistry of the main group elements, this chapter aims to introduce aspects of the structure and bonding of main group compounds. It is assumed that the reader has a basic understanding of atomic structure and bonding. We will start by discussing the concepts of ionization energy, electron affinity and electronegativity, which provide a framework for a brief overview of the chemistry of the main group elements, setting the scene for the detailed chemistry described in subsequent chapters. Valence shell electron pair repulsion theory (VSEPR), a powerful but simple theory for predicting and rationalizing the shapes adopted by main group compounds, is then discussed. The concept of molecular orbital theory will then be briefly covered.

More correctly, ionization energies and electron affinities should be referred to as **ionization enthalpies** and **electron attachment enthalpies**, respectively, though energies are commonly used.

1.2   Ionization Energy, Electron Affinity and Electronegativity

When an element forms a chemical compound, electrons are either lost, gained or shared with other atoms. These tendencies can be assessed by the parameters of **ionization energy (IE)**, **electron affinity (EA)** and **electronegativity**. Prediction of bond types as either ionic or covalent allows prediction of the chemical and physical properties of chemical substances.

IE refers to the loss of an electron from a gaseous atom or ion (equation 1.1). Successive loss of electrons from an atom becomes increasingly difficult (because the resulting positive ion holds on to its remaining electrons even more strongly), so, for example, third IEs are always higher than second IEs, which in turn are higher than first.

Ionization energies (kJ mol⁻¹)

Element	K	Al
First IE	+425	+584
Second IE	+3058	+1823
Third IE	+4418	+2751

$$M^{n+}_{(g)} \rightarrow M^{(n+1)+}_{(g)} + e^- \tag{1.1}$$

Going down a group, IEs decrease. Atoms increase in size and the electron to be removed is further from the nucleus; although the nuclear size is much increased, outer electrons are shielded by completed, filled, inner shells, so the **effective nuclear charge** felt by an outer electron is much less.

First IEs (kJ mol⁻¹)

Li	+526
Na	+502
K	+425
Rb	+409
Cs	+382

Crossing the Periodic Table, first IEs increase because extra protons are being added to the nucleus, and electrons are being added to the same electron shell (Figure 1.1). These electrons are not very efficient at screening each other from the nuclear charge, so they are attracted more strongly by the nucleus and are harder to ionize. Removal of an elec-

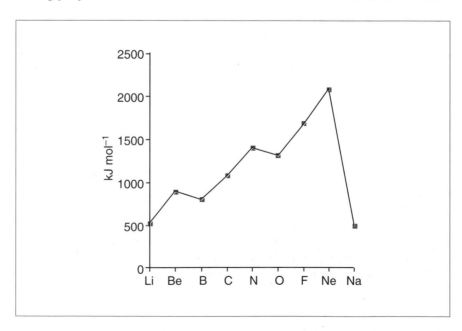

Figure 1.1 First ionization energies for the elements Li to Na

tron from a filled shell requires a large amount of energy: the first IE of neon is very high in comparison to the next element, sodium, where loss of an electron will leave a filled shell; Group 1 metals therefore have low first IEs but very high second IEs.

Superimposed on the general trend of increased IE with atomic number within a period, are 'kinks' at boron and oxygen. For beryllium, the 2s level is filled, so going to boron involves adding an electron to one of the 2p orbitals. Despite the increase in nuclear charge there is a decrease in IE because of the relatively efficient shielding of the 2p electron by the 2s electrons. At nitrogen, the three 2p orbitals each contain one electron (Hund's rule), so going to oxygen involves pairing an electron in one of the 2p orbitals. The two electrons in the same orbital repel each other, so the first IE of oxygen is lower than that of nitrogen, because loss of one of the paired electrons is assisted by electron–electron repulsions.

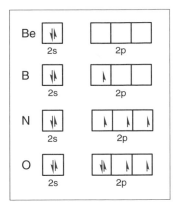

Worked Problem 1.1

Q Figure 1.2 shows the variation in second ionization energy for the elements Li to Na. Comment on the shape of the graph by comparison with first ionization energies of the same elements (Figure 1.1).

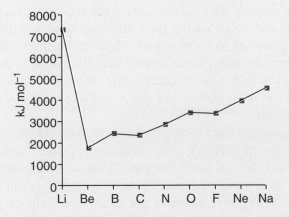

Figure 1.2 Second ionization eneergies for the elements Li to Na

A The same general trend is observed, with a general increase in second IE with increasing atomic number; for a given element the second IE is much higher than the first IE because it involves removal of an electron from a positive ion. The lowest second IE is for Be, since this involves the loss of the second 2s electron, leaving a filled $1s^2$ shell, and the highest second IE is for Li, since this involves loss of one of the 1s electrons (the sole 2s electron is lost in the first IE of Li). The kinks in the graph of second IE are similar of those in the first IE graph, except that the 'pattern' is shifted one element to the right. Thus, the decrease in first IE which occurs from N to O occurs in the second IE graph from O to F, because the process involves the loss of the same electrons.

Electron affinity (EA) is defined as the energy change on addition of an electron to a gaseous atom or ion (equation 1.2).

$$X^{n-}_{(g)} + e^- \rightarrow X^{(n+1)-}_{(g)} \qquad (1.2)$$

First electron affinities (kJ mol^{-1})

F	−322
Cl	−349
Br	−325
I	−295

EAs are most negative for elements in the top right of the Periodic Table (*i.e.* the halogens), while second and higher EAs are always positive because it is more difficult to add an electron to an already negatively charged ion. EAs also become positive on adding electrons to a new shell, so the first EAs of oxygen, fluorine and neon are −141, −322 and +29 kJ mol^{-1}, respectively. Trends in EA are generally less simple than trends in IE.

When we consider IEs and EAs together, elements in the bottom-left corner of the Periodic Table have low IEs and EAs and readily lose electrons to form cations, whereas elements in the top-right corner (the halogens, oxygen and sulfur) have high IEs and large negative EAs and readily gain electrons to form anions. Elements in the middle (particularly the lighter elements) have intermediate IEs and EAs and generally form covalent bonds in their compounds. However, unequal electron sharing results in polar bonds, and this is best discussed in terms of atom electronegativities.

Electronegativity refers to the tendency of an atom, *in a molecule*, to attract electrons to itself. A scale of electronegativity was devised by Linus Pauling, based on bond energies. While several other electronegativity scales have been developed, the one by Pauling is still widely used. The most electronegative elements are in the top right of the Periodic Table, with fluorine being the most electronegative with the maximum value of 4.0 on the Pauling scale.

Some electronegativities (Pauling scale)

F	4.0
Cl	3.0
O	3.5
N	3.0
S	2.5
C	2.5
H	2.1
B	2.0
Na	0.9

Electronegativity is a useful general parameter for predicting the general chemical behaviour of an element, and gives good indications of bond types. In general terms, two elements with a large electronegativity difference will tend to form ions, though smaller electronegativity differences are needed when one of the compounds is a highly electropositive metal (*e.g.* of Group 1). Two elements with similar and intermediate electronegativities (around 2.5) will tend to form covalent compounds. This is illustrated by C and H, which form an extensive range of covalently bonded organic compounds.

1.3   Periodic Trends among the Main Group Elements

The main group elements, and their chemical compounds, cover a wide range of bonding types, from ionic, through polymeric, to molecular. In this section we will survey the general features of the chemistry of the main group elements and selected compounds, using the variation in electronegativity of the elements as a qualitative tool for rationalizing the features. The compounds surveyed are the hydrides, oxides and chlorides, which are some of the most important compounds, and which illustrate the general features very well. More detailed discussion of individual compounds can be found within the appropriate chapter.

1.3.1 The Elements

It is noteworthy that the p block is the only part of the Periodic Table to contain non-metallic elements. There is a general trend from metallic elements at the bottom left of the Periodic Table (the s-block metals) to non-metallic elements at the top right of the Table (the halogens and noble gases) (Figure 1.3). This correlates very well with the electronegativities of the elements. Metals are good conductors of heat and electricity, and in solid metals the electrons are extensively delocalized over the whole material. Non-metallic elements are insulators and have no delocalized bonding, instead being formed from localized covalent bonds. In the centre of the p block, there arc also so-called **metalloid** elements such as boron and silicon, which show intermediate electronegativities; they also show relatively low electrical conductivity (compared to metals), but it increases with temperature.

The change in properties is nicely illustrated by looking at the first long period, Na to Ar. Na and Mg are both electropositive metals; the next element, aluminium, is a metal, but shows several characteristics of non-metals in forming many covalent compounds. In Group 14, silicon is a metalloid, the element being a semiconductor, and has compounds which show characteristics of both metal compounds and non-metal compounds. By the time we get to phosphorus in Group 15, we are truly in the domain of non-metals; phosphorus exists in several elemental forms, all of which contain covalent P–P bonds. In Groups 16 (sulfur) and 17 (chlorine) the elements are also true non-metals, sulfur existing as covalent S_8 rings (and other forms), and chlorine forming diatomic, covalently bonded molecules. Argon exists as a monoatomic gas under ambient conditions, and does not participate in chemical bonding owing

Main group elements can be roughly classified as metals with an electronegativity <2, and as non-metals with electronegativities >2.2.

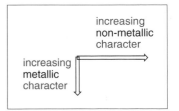

Figure 1.3 Variation in metallic and non-metallic character in the Periodic Table

to the filled valence shell of electrons, and high ionization energy.

Going down any of the main groups, elements become more metallic in character, paralleled by a decrease in electronegativity.

1.3.2 Main Group Element Hydrides

The chemistry of hydrides is discussed in Section 2.4.

The properties of the main group element hydrides range from ionic (for the s-block metals, with the exception of beryllium), through polymeric (AlH_3), to molecular covalent hydrides for the elements of Groups 14–17.

In Groups 1 and 2, the metals are less electronegative than hydrogen (Pauling scale electronegativities: Na 0.9, H 2.1), so the bonding in the hydrides of these metals is predominantly ionic, as M^+H^-; these hydrides react violently with water, generating H_2 gas. For boron and beryllium, the electronegativity diference between the element and hydrogen is small. BeH_2 is covalent and boron hydrides are covalent clusters. In Group 14 the hydrides are all covalent molecular species, typified by CH_4. Continuing across the Periodic Table, to Groups 15, 16 and 17, the hydrides are all molecular covalent species, with acidity in aqueous solution increasing on moving to the right, as the electronegativity difference between the element and hydrogen increases and the H–X bond becomes more polarized: $H^{\delta+}-X^{\delta-}$. This has a marked effect on the physical properties (*e.g.* boiling points) of the hydrides of electronegative elements, as described in Section 2.6.1.

Worked Problem 1.2

Q Predict the properties of the hydrides formed by elements with electronegativities of (a) 0.9 and (b) 3.5.

A (a) An element with an electronegativity of 0.9 is a metal; it will probably form an ionic hydride which will react with water, giving hydrogen and a basic solution of the hydroxide. (b) This element is a non-metal, and the hydride will be covalent with a polar H–X bond and will dissolve in water, probably giving a neutral to acidic solution.

1.3.3 Main Group Element Chlorides

Like the hydrides, the properties of the chlorides follow a broadly similar pattern, with chlorides of metals being ionic and of non-metals being covalent molecular in structure. Thus, for the Group 1 and 2 metals (except beryllium) the chlorides are ionic solids which form neutral solutions in water. The chlorides of small, highly polarizing metal ions such as beryllium, aluminium, gallium and some other elements are polymeric in the solid state. The majority of the chlorides of the Groups 14 and 15 elements, and BCl_3, are molecular covalent species. The chlorides of the p-block elements and beryllium generally give acid solutions in water, because they react with it rather than simply dissolving. It is noteworthy that CCl_4, unlike $SiCl_4$, does not react with water to give an acidic solution; this is purely a kinetic effect, and is discussed in Section 6.5.1.

1.3.4 Main Group Element Oxides

For main group oxides, there is a similar trend from ionic oxides for the bottom left elements, through polymeric oxides in the centre (many of which are **amphoteric**), to molecular covalent oxides for the elements of higher electronegativity on the right-hand side of the p block.

An amphoteric oxide dissolves in both acidic and basic (alkaline) solutions.

Oxygen is the second most electronegative element, so in combination with (low electronegativity) Group 1 and 2 metals at the left-hand side of the Periodic Table, the resulting oxides are ionic. Examples include Na_2O and CaO. Such oxides are basic oxides, giving highly alkaline solutions in water (equation 1.3). On moving to the right, to Group 13, the oxides such as B_2O_3 and Al_2O_3 are polymeric and Al_2O_3 is amphoteric. In Group 14 the oxides of the lightest element, carbon, such as CO and CO_2, are molecular oxides; in marked contrast, SiO_2 is a polymeric oxide. CO_2 is an example of an acidic oxide, since it dissolves in water giving an acidic solution (see Section 6.7.1). In Groups 15 and 16 the oxides of nitrogen are all molecular covalent species, many of which are acidic, while those of sulfur (SO_2 and SO_3) are both acidic oxides (equation 1.4). Likewise, in Group 17, and for xenon in Group 18, the oxides are molecular species.

$$Na_2O_{(s)} + H_2O_{(l)} \rightarrow 2Na^+_{(aq)} + 2OH^-_{(aq)} \qquad (1.3)$$

$$SO_{3(s)} + H_2O_{(l)} \rightarrow 2H^+_{(aq)} + SO_4^{2-}_{(aq)} \qquad (1.4)$$

1.4   Valence Shell Electron Pair Repulsion Theory

1.4.1   Introduction

A cursory inspection of the compounds formed by the p-block elements in subsequent chapters reveals that many structures are observed. Even for a certain fixed number of groups around a central atom, there are often different geometrical ways of arranging these; for example, five-coordinate species may be either trigonal bipyramidal or square pyramidal.

The shape of a molecule is described by the spatial arrangement of the atoms, disregarding the positions of any **lone pairs** (often called **non-bonding pairs**). Thus, ammonia, NH_3, is pyramidal, even though the N has four pairs of electrons in its valence shell, arranged approximately tetrahedrally.

The simplest and most widely practised method for shape prediction is **valence shell electron pair repulsion theory (VSEPR)**, originally developed in the 1960s, and recently redeveloped by Gillespie.[1] Prediction (or ideally knowledge) of molecular shapes is important for prediction of properties dependent on molecular shape, for example boiling points. The knowledge of bond polarity, determined using the concept of electronegativity, is also important.

1.4.2 Basic Principles of VSEPR

The basic premise of VSEPR is that *pairs of electrons* in the *valence shell* of the central atom of a molecule *repel* each other and take up positions as far apart as possible. The core electrons, which cannot easily be polarized, are conveniently ignored. The shape of a molecule thus condenses to a simple geometrical 'points-on-a-sphere' model, and the basic shapes adopted by molecules with between two and six pairs of electrons on the central atom are given in Table 1.1. In order to predict the *basic* shape of a molecule or ion by VSEPR, the general procedure in Box 1.1 should be followed. It is important to note that the shape of a molecule or ion can be predicted without knowing anything about the bonding in that species (see Section 1.5).

Table 1.1 Shapes of molecules and ions

Number of central atom electron pairs	Bonding pairs	Non-bonding pairs	Shape	Example
2	2	0	Linear	$BeCl_2$
3	3	0	Triangular	BF_3
3	2	1	Bent	$SnCl_2$
4	4	0	Tetrahedral	CCl_4
4	3	1	Pyramidal	NH_3
4	2	2	Bent	H_2O
5	5	0	Trigonal bipyramidal (tbp)	PF_5
5	4	1	Pseudo-tbp	BrF_4^+, SF_4
5	3	2	T-shaped	BrF_3
5	2	3	Linear	XeF_2
6	6	0	Octahedral	SF_6, PF_6^-
6	5	1	Square pyramidal	IF_5
6	4	2	Square planar	XeF_4, IF_4^-

Box 1.1 Predicting the Shapes of Molecules and Ions using VSEPR

1. Draw a simplified Lewis structure for the molecule, noting the presence of any formal double, triple and dative bonds. The presence of a charge on the central atom should also be identified; if the charge is not on the central atom, then it can be ignored.
2. Count electrons on the central atom, taking the central atom as a neutral atom. An atom in Group 14 will have four valence electrons, an atom in Group 15 will have five, *etc.*
3. Add one electron for every atom σ-bonded to the central atom, but two electrons from any dative bonds from other atoms.
4. Subtract one electron for every π-bond present involving the central atom.
5. If the central atom has a positive charge, subtract the appropriate number of electrons, or add the appropriate number of electrons if it has a negative charge.
6. Thus obtain the number of *electron pairs* on the central atom. By consulting Table 1.1, the basic shape of the molecule can then be determined.

Worked Problem 1.3

Q Predict the shape of carbon tetrabromide, CBr_4.

A Carbon is in Group 14, and has four valence electrons; σ-bonds to four Br atoms contribute a total of four electrons. The central atom has no charge, so there are eight electrons, or four pairs, on the carbon. There are four bonding pairs and four Br atoms to be bonded, so the shape is therefore a regular tetrahedron.

The bond angles in a regular tetrahedron are all 109.5°.

Lewis structure:

Shape:

(tetrahedral)

Box 1.2 Hybridization

The concept of hybridization is used to combine atomic orbitals on an atom to generate suitable orbitals which point in the directions required. The hybridization of a carbon 2s-orbital with three carbon 2p-orbitals generates four equivalent sp^3 hybrid orbitals, which point towards the vertices of a regular tetrahedron, so the carbon in CBr_4 can be considered to be sp^3 hybridized. Similarly, for bond angles of 120° or 180°, sp^2 or sp hybrids can be used, while in a trigonal bipyramid sp^3d hybrids, and in an octahedron sp^3d^2 hybrids, point in the required directions, and so can accommodate the electron pairs identified in the VSEPR analysis.

The role of hybridization must be kept in perspective: it is a mathematical means of generating orbitals pointing in the required directions.

1.4.3 Molecules containing Non-bonding Pairs of Electrons

A bonding pair is shared by *two atoms* whereas a lone pair is only held by *one atom*. A lone pair therefore occupies more space in the valence shell of the atom to which it belongs, and it will exert a larger repulsive influence on the other pairs of electrons on that atom. Therefore, in general terms:

lone pair–lone pair repulsion > lone pair–bonding pair repulsion > bonding pair–bonding pair repulsion

This is best illustrated by two worked examples: ammonia, NH_3, and water, H_2O.

The experimentally determined shape of ammonia

Worked Problem 1.4

Q Predict the shape of the ammonia molecule.

A Nitrogen has five valence electrons; three σ-bonds to H contribute three electrons. Thus there is a total of eight electrons, or four pairs, and therefore the *distribution of electron pairs* is tetrahedral. Since there are four pairs of electrons and three hydrogens, there is a lone pair. Hence the *shape of the molecule* is described as **trigonal pyramidal**. The lone pair will occupy more space in the valence shell of N, so the N–H bonds will be pushed towards each other slightly, decreasing the H–N–H bond angle from that of a regular tetrahedron.

Worked Problem 1.5

Q Predict the shape of water.

A The oxygen of water also has four pairs of valence electrons. The geometry of the water molecule is now described as bent, and since there are two lone pairs on oxygen, these try to repel each other slightly more, and so the hydrogens are pushed slightly closer, reducing the H–O–H angle from that in a regular tetrahedron even further.

The experimentally determined shape of water

1.4.4 Molecules with Multiple Bonds

A double bond contains both σ and π components. For the purposes of VSEPR, both can be considered to point in the same direction, and we can thus treat a double or triple bond as one 'superpair' of electrons, the effect of which is rather similar to that of a lone pair. As an example, consider the molecule COF_2. The molecule will be trigonal, but the greater space occupied by the double bond will make the fluorine atoms move closer together, decreasing the F–C–F bond angle.

1.4.5 Molecules with Five Electron Pairs

In a tetrahedron or octahedron, all of the vertices are identical; however, this is not the case for a trigonal bipyramid, where there are two different types of vertex: **axial** and **equatorial**. This is best illustrated by an example, that of PCl_5 in the gas phase, shown in Figure 1.4. The molecule is a regular trigonal bipyramid.

The P–Cl(axial) bond length (214 pm) of PCl_5 is slightly longer than the P–Cl(equatorial) bond length (202 pm) (where 1 pm = 1 picometre = 10^{-12} m).

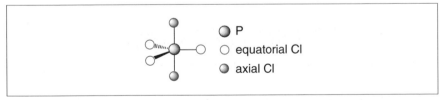

○ P
○ equatorial Cl
● axial Cl

Figure 1.4 The structure of gas-phase PCl_5, showing the presence of Cl atoms in axial and equatorial positions

This has implications for molecules which contain five electron pairs, with one or more lone pairs, since there will be a choice of putting the lone pair(s) in axial or equatorial positions. However, on considering the various repulsions in such species, it can be concluded that: *lone pairs of electrons, or multiple bonds, always adopt equatorial positions in trigonal bipyramids, owing to the greater space occupied in the valence shell of the central atom.* This can be illustrated by species such as SF_4, BrF_4^+, ClF_3 and XeF_2.

Worked Problem 1.6

Q Predict the shape of SF_4.

A S has six valence electrons and the four fluorines contribute four electrons, giving 10 electrons or five pairs. The experimentally determined shape of SF_4 is a 'see-saw' shape, sometimes called a *pseudo* trigonal bipyramid (Figure 1.5). The effect of the sulfur lone pair can be clearly seen, pushing the fluorines towards each other. The molecule SOF_4 has a very similar shape, with the (double bonded) oxygen replacing the sulfur lone pair.

Figure 1.5 The structure of gas-phase SF_4, showing the presence of F atoms in axial and equatorial positions. The strong repulsing effect of the equatorial lone pair is clearly shown

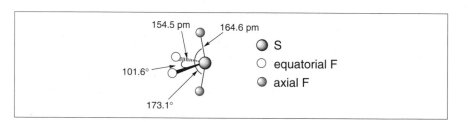

154.5 pm   164.6 pm

101.6°

173.1°

○ S
○ equatorial F
◉ axial F

1.4.6 Molecules and Ions with Seven or More Electron Pairs

When seven electron pairs are present in the valence shell of the central atom, the shape is more difficult to predict; there are often a number of different arrangements with similar energies. The three most important regular shapes are the monocapped octahedron, the monocapped trigonal prism and the pentagonal bipyramid, shown in Figure 1.6.

Figure 1.6 Common shapes adopted by seven-electron pair species: (a) monocapped octahedron; (b) monocapped trigonal prism; (c) pentagonal bipyramid

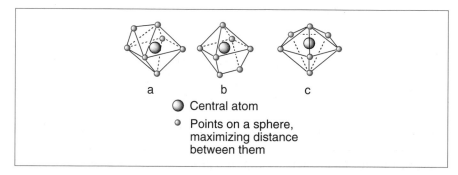

a   b   c

◉ Central atom
◉ Points on a sphere, maximizing distance between them

In certain cases, for the heavy p-block elements such as selenium, tellurium, bromine and bismuth, lone pairs occupy spherical s-orbitals, which do not influence the geometry of the species. As an example,

$TeCl_6^{2-}$ has one lone pair and six Cl atoms, but it is a regular octahedron; $SeCl_6^{2-}$ and BrF_6^- are also octahedral for the same reason.

An example of a species with eight electron pairs is XeF_8^{2-}, which has a square antiprism shape, shown in Figure 1.7.

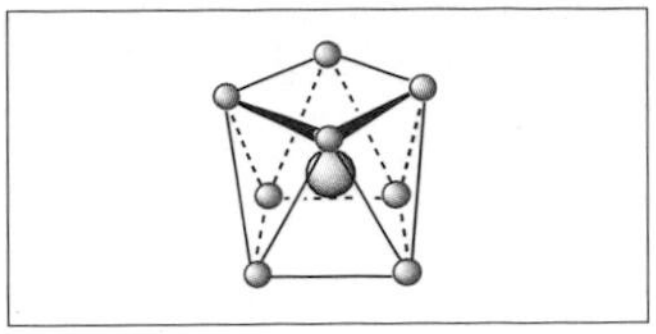

Figure 1.7 The square antiprismatic arrangement of eight fluorines around a central xenon atom in the ion XeF_8^{2-}. (A square antiprism can be derived from a cube by rotation of one of the square faces through 45°)

1.4.7 Resonance

When there are two or more **resonance forms** for a molecule or ion, it is essential that these are considered before predictions of bond angles are made. This can be illustrated by the example of the carbonate ion, CO_3^{2-}. This ion has six electrons (three pairs) around the carbon atom, and so is trigonal. As shown in Scheme 1.1, there are three resonance hybrids of the carbonate ion, so the true structure is a blend of all three, structure **1.1**, and all of the O–C–O bond angles are exactly 120°. If *one* of the resonance forms was taken in isolation, with a C=O double bond and two C–O single bonds, it would be (incorrectly) predicted that the O=C–O bond angles would be >120° and the O–C–O bond angle <120°. The importance of considering all resonance forms before predicting bond angles can be clearly seen.

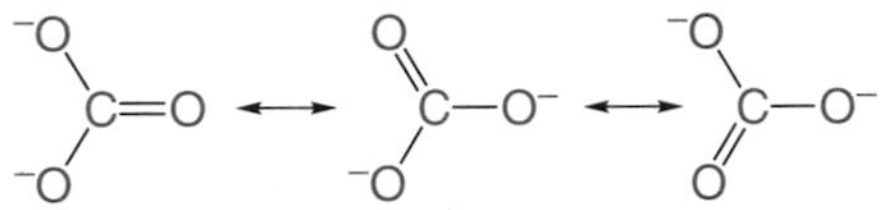

Scheme 1.1

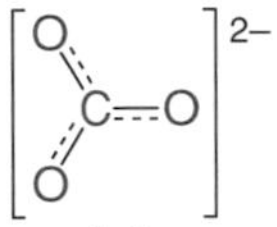

1.1

1.4.8 Dative Bonds

A dative bond is fundamentally identical to a 'normal' two-electron covalent bond except that in our electron 'book-keeping' we consider both electrons in the bond to originate from the same atom.

Worked Problem 1.7

Q Predict the shape of $Et_2O{\rightarrow}BF_3$ (Section 5.5.1).

A In this species, both electrons in the O–B bond come from the oxygen. There are two central atoms (O and B): for the B, we add two electrons to the count, both coming from O, whereas for the O, zero electrons are provided from the B. Thus:

For B:
B has three valence electrons
3F provides three electrons

Dative bond provides two
electrons

Total eight electrons
(four pairs)

For O:
O has six valence electrons
Two Et groups provide two
electrons
B provides 0 electrons
(dative bond is from O to B)

Total eight electrons
(four pairs)

Thus the geometry about B is tetrahedral, while around O it is trigonal pyramidal (with one non-bonding pair).

The structure of $Et_2O–BF_3$

1.4.9 Atom Electronegativities

Electronegativity: see Section 1.2.

In an A–X bond between atoms A and X, as the atom X becomes more electronegative, the bonding pair occupies less space in the valence shell of atom A. In practice, this means that in a related series of compounds, bond angles of the type F–A–F are typically smaller than Cl–A–Cl or Br–A–Br angles, as illustrated in Figure 1.8. The greater size of a Cl atom compared to an F atom also contributes to the widening of the Cl–A–Cl angle.

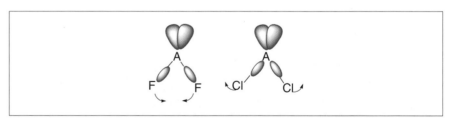

Figure 1.8 Effect of atom electronegativity on bond angle

Worked Problem 1.8

Q Explain why the X–P–X bond angles for the series of POX_3 molecules decrease from X = Br (104.1°) to X = Cl (103.3°) to X = F (101.3°).

A Fluorine is the most electronegative halogen, so it will draw electron density in the P–F bond away from the P atom; repulsion of the P–F bonding pairs will be less than repulsion of P–Cl and P–Br bonding pairs, so the F–P–F bond angle will be the smallest.

Worked Problem 1.9

Q Which of H_2O and F_2O will have the larger X–O–X bond angle?

A F is more electronegative than H; therefore the space occupied by the O–H bonding pair in the O valence shell will be greater. Hence, H_2O will have the larger bond angle

The H–O–H and F–O–F bond angles in H_2O and F_2O are $104.5°$ and $103.1°$, respectively.

CH_4 and CF_4 have regular tetrahedral shapes, but CF_2H_2 is a distorted tetrahedron with the F–C–F bond angle smaller than the H–C–H angle. Following the same logic as in worked problems 1.8 and 1.9, this is because the electronegative fluorines attract electrons in the C–F bonds, decreasing the space they occupy in carbon's valence shell, causing the F–C–F angle to decrease.

Box 1.3 Bent's Rule

Bent's rule[2] states: *More electronegative substituents 'prefer' hybrid orbitals having less s-character, and more electropositive substituents 'prefer' orbitals having more s-character.*

  The bond angles in CH_4, CF_4 and CH_2F_2 can be explained using Bent's rule. While a carbon atom in CH_4 or CF_4 uses four identical sp^3 hybrids in bonding, in CF_2H_2 the hybrids used are not identical. The C–F bonds are formed from sp^{3+x} hybrids, with slightly more p-character and less s-character than an sp^3 hybrid, and the hydrogens are bonded by sp^{3-x} hybrids, with slightly less p-character and slightly more s-character. Increasing the amount of p-character in the C–F bonds decreases the F–C–F bond angle, because for bonding by pure p-orbitals the bond angle would be decreased to $90°$.

Recall that the three p-orbitals, p_x, p_y and p_z, point at $90°$ to each other along the x, y and z axes.

1.5   Molecular Orbital Theory

In molecular orbital (MO) theory, rather than having localized orbitals which form bonds between pairs of atoms, we construct molecular orbitals which extend over all atoms in a molecule. Space prevents more than a brief summary of the MO treatment of bonding, and in this section it is intended to illustrate the application of MO theory to main group molecules. The discussion will be illustrated by considering the properties of various O_2 species (O_2^+, O_2, O_2^- and O_2^{2-}).

  We will start by looking at a very simple molecule: dihydrogen, H_2. Each H atom has a 1s atomic orbital (AO) available for bonding, and these can interact in two ways. In-phase interaction gives a bonding MO, σ, while out-of-phase interaction gives an antibonding MO, σ*, as shown in Figure 1.9. The bonding MO is symmetrical about the centre of the molecule, and there is an increase in electron density in the internuclear region compared to the two H 1s orbitals; the bonding MO σ therefore has a lower energy than the energy of the hydrogen 1s atomic orbital. In contrast, the antibonding MO σ* has a decrease in electron density

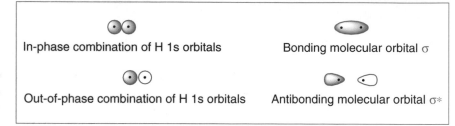

In-phase combination of H 1s orbitals   Bonding molecular orbital σ

Out-of-phase combination of H 1s orbitals   Antibonding molecular orbital σ*

Figure 1.9 Molecular orbitals of H_2. The small black dots indicate the nucleus; these are omitted on subsequent diagrams

An antibonding MO is denoted by the * symbol.

in the internuclear region (when occupied by electrons), and is higher in energy than the 1s atomic orbitals. Figure 1.10 shows an energy level diagram for the H_2 molecule. The two electrons (one from each hydrogen atom) enter the bonding MO σ, corresponding to a net H–H single bond. If two more electrons are added to H_2, the antibonding MO σ* becomes filled. In this case, equal numbers of electrons are present in bonding and antibonding MOs, there is no net bonding, and H_2^{2-} is unfavoured relative to its dissociation product, two H^- (hydride) anions.

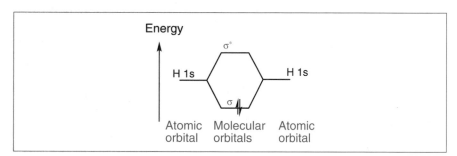

Figure 1.10 MO energy level diagram for H_2

Degenerate orbitals are ones with the same energy.

In dioxygen species O_2, the MO scheme is slightly more complex because each O atom has both 2s and 2p AOs available for bonding. The MO energy level diagram is shown in Figure 1.11. Similar to H_2, the 2s AOs of the oxygen atoms interact to form bonding $σ_1$ and antibonding $σ_2$* MOs. The p_z orbitals point towards each other (the molecular axis is defined as the z axis) and can interact to form σ bonding ($σ_3$) and antibonding ($σ_4$*) MOs. The p_x orbitals on each oxygen can interact in a side-on manner, to form π bonding ($π_1$) and antibonding ($π_2$*) MOs, respectively. In the same way, the p_y orbitals interact to form π bonding ($π_1$) and antibonding ($π_2$*) MOs which are degenerate with the p_x-derived π MOs. The MO diagram in Figure 1.11 can be used to describe the bonding in O_2^+, O_2, O_2^- and O_2^{2-}. As with atoms, electrons enter MOs from the lowest energy levels first, and if there are degenerate orbitals, electrons initially singly occupy each with parallel spins.

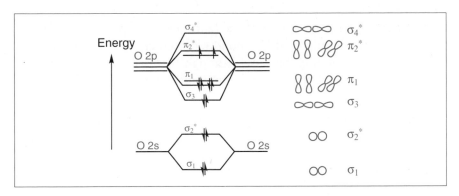

Figure 1.11 MO energy level diagram for dioxygen, O_2

For O_2 it can be seen that the π_2^* MO contains two unpaired electrons, and hence O_2 is predicted to be paramagnetic and attracted by a magnetic field. This fits very well with the physical properties of O_2, which is indeed paramagnetic. It is important to note that a simple valence bond description of O_2 (as O=O) does not predict any unpaired electrons, and this therefore represents one of the major triumphs of MO theory. The total **bond order** for O_2 from the MO diagram is 2, in accordance with the simple valence bond description of O=O.

> Bond order is defined as $\frac{1}{2} \times$ [electrons in bonding MOs minus electrons in antibonding MOs].

The oxygenyl cation, O_2^+, has one less electron in the antibonding π_2^* MO, so the bond order is 2.5. Similarly, O_2^- and O_2^{2-} have respectively one and two more electrons in π_2^*, so the bond orders are 1.5 and 1, respectively. The predicted bond orders correlate very well with the experimental bond lengths, with a larger bond order giving a shortened bond, as shown in Table 1.2. Similarly, O_2^+, O_2 and O_2^- contain unpaired electrons and are paramagnetic, while O_2^{2-} has no unpaired electrons and is diamagnetic.

Table 1.2 Bond lengths of some dioxygen species[a]

Species	Name	Bond order	Bond length (pm)
O_2^+	Oxygenyl	2.5	112.3
O_2	Dioxygen	2.0	120.7
O_2^-	Superoxide	1.5	128
O_2^{2-}	Peroxide	1.0	149

[a] Refer to Chapter 8 for further details on the chemistry of these species.

Worked Problem 1.10

Q Identify the following orbital interactions as bonding, non-bonding or antibonding:

A (a) This interaction has in-phase (end-on) overlap of an s-orbital with a p-orbital, and so it is a bonding interaction. (b) This interaction is non-bonding, because although there is a bonding interaction, it is cancelled out by an equal antibonding interaction.

(a)

(b)

Summary of Key Points

1. The *overall chemistry* of the main group elements can be classified using the principles of ionization energy and electronegativity. Elements at the bottom left of the main group show strong metallic properties, forming basic oxides and hydrides and neutral halides. Non-metals in the top right of the p-block have high ionization energies and electron affinities, and form acidic oxides, hydrides and halides.

2. Valence shell electron pair repulsion theory (VSEPR) is a simple, powerful method for predicting the *molecular shapes* of main group species. While VSEPR can be used to predict shapes, it says nothing about the bonding in main group compounds.

3. Molecular orbital (MO) theory can be used to predict the *bonding* and properties of many species, and the discussion has centred around dioxygen species as an example.

Problems

1.1. Which of the following species will have the larger first ionization energy: (a) Li or Be; (b) N or O; (c) C or N; (d) Se or Se^+; (e) K or Rb.

1.2. Predict the shapes of the following molecules or ions: (a) SF_6; (b) SeF_2; (c) HCO_3^-; (d) $XeOF_4$; (e) PF_3Cl_2; (f) $[SF_2Cl]^+$; (g) $[S_2O_4]^{2-}$.

1.3. Which of the following molecules has a bond angle greater than $109.5°$: (a) SF_2; (b) CF_4; (c) BF_3; (d) PF_3; (e) H_2S.

1.4. Using the MO diagram in Figure 1.9, predict the bond order of the species O_2^{2+}. With which common molecule is it isoelectronic?

1.5. Classify the following orbital interactions as bonding, nonbonding or antibonding:

(a)

(b)

References

1. R. J. Gillespie, *Chem. Soc. Rev.*, 1992, **21**, 59.
2. H. A. Bent, *J. Chem. Educ.*, 1960, **37**, 616; *Chem. Rev.*, 1961, **61**, 275; J. E. Huheey, E. A. Keiter and R. L. Keiter, *Inorganic Chemistry*, 4th edn., HarperCollins, New York, 1993.

Further Reading

M. J. Winter, *Chemical Bonding*, Oxford University Press, Oxford, 1994.
Electron domains and the VSEPR model of molecular geometry, R. J. Gillespie and E. A. Robinson, *Angew. Chem., Int. Ed. Engl.*, 1996, **35**, 495.
Multiple bonds and the VSEPR model, R. J. Gillespie, *J. Chem. Educ.*, 1992, **69**, 116.
Lewis structures, formal charge and oxidation numbers, J. E. Packer and S. D. Woodgate, *J. Chem. Educ.*, 1991, **68**, 456.
Teaching a model for writing Lewis structures, J. Q. Pardo, *J. Chem. Educ.*, 1989, **66**, 456.

2
The Chemistry of Hydrogen

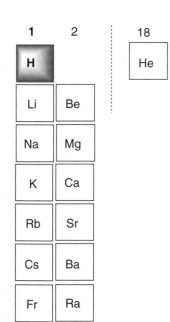

Aims

By the end of this chapter you should understand:

*   The unique position of hydrogen in the Periodic Table
*   The formation of different types of hydrides
*   Hydrogen bonding

2.1   Introduction

Hydrogen is the simplest element in the Periodic Table, yet its chemistry is among the most diverse. It does not satifactorily belong to any single group of the Periodic Table, but has an electron configuration (s^1) similar to those of the alkali metals, and is sometimes shown at the top of this group.

Hydrogen has, in simplistic terms, three different ways in which it forms chemical compounds:

A high charge-density ion is one with a high charge-to-radius ratio (with a small size and/or a high charge)

The term 'hydride' is used to refer to covalent and interstitial hydrogen compounds as well as the H⁻ anion.

*   It can *lose* its single electron to form a proton, H^+, which has no chemical existence on its own, but is always solvated, owing to its extremely high charge density.
*   It can *share* an electron through formation of covalent bonds with many other elements.
*   It can *gain* an electron, to form the **hydride ion**, H^-, which has a s^2 electronic configuration isoelectronic with the noble gas helium.

2.2 The Element

2.2.1 Occurrence and Manufacture

Hydrogen is the most abundant element in the universe, followed by helium. However, in Nature, free dihydrogen (H_2) is very reactive, and the majority of the element exists in chemical combination with other elements, particularly as water in the oceans and in hydrated minerals, as well as organic compounds. H_2 can be prepared by several methods:

- Electrolysis, either of water itself (equation 2.1), or the electrolysis of brines (using Hg as an electrode) to produce chlorine, where elemental hydrogen is formed as a by-product (equation 2.2).

$$2H_2O_{(l)} \rightarrow 2H_{2(g)} + O_{2(g)} \quad\quad (2.1)$$

$$2NaCl_{(aq)} + 2Hg \rightarrow 2NaHg_{(l)} + Cl_{2(g)}$$

then

$$2NaHg_{(l)} + H_2O_{(l)} \rightarrow 2NaOH_{(aq)} + H_{2(g)} + 2Hg_{(l)} \quad (2.2)$$

- Reforming of hydrocarbons: hydrocarbons react with steam over a nickel catalyst at around 800 °C, *e.g.* equation 2.3 for the reforming of methane. This is the main industrial method for H_2 manufacture.

$$CH_4 + H_2O \rightarrow CO + 3H_2 \quad\quad (2.3)$$

The CO–H_2 mixture is often called **synthesis gas** because it is used to synthesize other compounds such as methanol.

- Thermal cracking of hydrocarbons: hydrogen is produced as a by-product when large hydrocarbons (alkanes) are thermally broken up into smaller alkenes.
- Hydrolysis of an ionic hydride, such as NaH or CaH_2 (equation 2.4).

$$CaH_2 + 2H_2O \rightarrow Ca(OH)_2 + 2H_2 \quad\quad (2.4)$$

Worked Problem 2.1

Q An alternative method for the synthesis of hydrogen in the laboratory is to react electropositive metals with acids (including the weak acid water). Give two examples of metals which react in this way, and write balanced equations for the reactions.

A Alkali (Group 1) metals, calcium, strontium and barium react with water, *e.g.*

$$2Na + 2H_2O \rightarrow 2NaOH + H_2$$

Less reactive metals such as magnesium, zinc and tin dissolve in acids such as HCl:

$$Mg + 2HCl \rightarrow MgCl_2 + H_2$$

Worked Problem 2.2

Q How much energy is produced by combustion of 4 g of hydrogen (ΔH for $H_{2(g)} + 1/2O_{2(g)} \rightarrow H_2O_{(g)}$ is -242 kJ mol^{-1})? What are the advantages of using hydrogen as a fuel?

A 4 g of hydrogen is 2 moles, so 484 kJ of energy would be produced. Hydrogen is an energy-rich fuel, and the combustion product is simply water, so hydrogen is a very attractive fuel, and is increasingly being used in transportation systems worldwide.

2.2.2 Isotopes of Hydrogen

There are three isotopes of hydrogen, listed in Table 2.1. The isotope ^1H is the dominant one; however, natural hydrogen also contains around 0.02% deuterium. All three isotopes are chemically identical, except that the different isotopes react at different rates. This difference in reaction rate is used in the production of deuterium (D_2). An example is the electrolysis of water (equation 2.1), where the hydrogen gas produced is enriched in ^1H and the residual water is enriched in the heavier, and more slowly reacting, D.

Table 2.1 The isotopes of hydrogen

Isotope	Name (symbol)	Protons	Neutrons	Stability
^1H	Protium or hydrogen (H)	1	0	Stable
^2H	Deuterium (D)	1	1	Stable
^3H	Tritium (T)	1	2	Radioactive (half-life 12.26 years)

Tritium is formed by cosmic ray bombardment of ^{14}N in the upper atmosphere, or in a nuclear reactor, where Li is bombarded with neutrons (n) (equation 2.5).

$$^6Li + n \rightarrow {}^3H + {}^4He \tag{2.5}$$

D_2O and T_2O are the common commercial sources of deuterium and tritium. Deuterated and tritiated compounds are produced using straightforward reactions, such as the preparation of deuterobenzene (C_6D_6), D_2SO_4 or ND_3 (equations 2.6–2.8).

$$3CaC_2 + 6D_2O \rightarrow 3DC \equiv CD \ [+3Ca(OD)_2] \xrightarrow{\text{catalyst}} C_6D_6 \quad (2.6)$$

$$SO_3 + D_2O \rightarrow D_2SO_4 \quad (2.7)$$

$$Li_3N + 3D_2O \rightarrow ND_3 + 3LiOD \quad (2.8)$$

Worked Problem 2.3

Q How would you prepare (a) D_3PO_4 and (b) $DCl_{(g)}$ starting from D_2O?

A (a) $P_4O_{10} + 6D_2O \rightarrow 4D_3PO_4$
(b) Prepare D_2SO_4 as in equation 2.7; then:
$D_2SO_{4(l)} + NaCl_{(s)} \rightarrow DCl_{(g)} + NaDSO_{4(s)}$
Alternatively, electrolysis of D_2O will give $D_{2(g)}$, which reacts with $Cl_{2(g)}$:
$D_{2(g)} + Cl_{2(g)} \rightarrow 2DCl_{(g)}$

2.3 The Chemistry of Hydrogen

The bond dissociation energy of H_2 is high, $+436$ kJ mol^{-1}, and so it is relatively unreactive at room temperature. However, in the presence of a catalyst, or at elevated temperatures, hydrogen is very reactive towards most elements. Atomic hydrogen (produced by passing an electrical discharge through a low pressure of H_2) is very reactive, because the strength of the H–H bond has been overcome by a non-chemical method.

Hydrogen combines with nitrogen, oxygen and the halogens directly, to give covalent hydrides (Section 2.4.2); the reactions often require initiation. Elemental hydrogen is a good reducing agent, and will reduce many metal oxides to the metal (plus water), and will hydrogenate many unsaturated organic compounds containing C≡C, C=C and C=O bonds to their saturated analogues (hydrocarbons and alcohols).

Some other bond dissociation energies (kJ mol^{-1}) for comparison:
Cl–Cl	+242
C–H	+414
N–H	+391

A well-known application is the partial hydrogenation of vegetable oil to give margarine, using a nickel catalyst.

2.4   Hydrides

The chemical combination of an element with hydrogen produces compounds called **hydrides**. Generally, the p-block elements produce covalent hydrides, the s-block metals (except Be and Mg) produce ionic hydrides, while the transition metals and lanthanides produce metallic hydrides, which have the appearance of metals, have electrical conductivity, but which (unlike metals) are brittle. Going across a period, there is a transition from ionic (*e.g.* NaH), to covalent polymeric (*e.g.* AlH_3), to covalent molecular hydrides (*e.g.* H_2S, PH_3). The different types of hydrides will be discussed in turn.

2.4.1 Ionic Hydrides

Reaction of Group 1 and 2 metals with hydrogen gas produces colourless hydrides containing the H^- ion, with an s^2 electronic configuration. When the electronegativity of the metal is less than about 1.2, then the hydride is ionic.

In the formation of ionic hydrides, hydrogen is mimicking the chemistry of the halogens, which are also one electron short of a noble gas electronic configuration. The formation of Cl^- (equation 2.9) is more favourable than H^- (equation 2.10) owing to the large **electron affinity** of Cl (-369 kJ mol^{-1}) versus H (-69 kJ mol^{-1}), and the weaker Cl–Cl bond compared to H–H.

$$0.5Cl_2 + e^- \xrightarrow{+121} Cl_{(g)} \xrightarrow{-369\,kJ\,mol^{-1}} Cl^-_{(g)} \quad \text{overall } \Delta H = -248\,kJ\,mol^{-1}$$
(2.9)

$$0.5H_2 + e^- \xrightarrow{+217.5} H_{(g)} \xrightarrow{-69\,kJ\,mol^{-1}} H^-_{(g)} \quad \text{overall } \Delta H = +148.5\,kJ\,mol^{-1}$$
(2.10)

Equation 2.11 is essentially $H^+ + H^- \to H_2$.

Unlike halide ions, which are stable in water, hydride ions are readily hydrolysed (equation 2.11).

$$H^- + H_2O \to H_2 + OH^-$$
(2.11)

The chemistry of Na[BH$_4$] and Li[AlH$_4$] is described in Section 5.6.

The hydrides of the alkali metals (Section 3.7) and of calcium, strontium and barium (Section 4.6) are essentially ionic materials. In contrast, hydrides of aluminium (Section 5.6.2), magnesium and beryllium (Section 4.6) are polymeric solids.

2.4.2 Covalent Hydrides

Hydrogen is able to form covalent bonds with other atoms by sharing its 1s electron; elements with electronegativities as low as 1.5 form hydrides with mainly covalent character.

There is a full range of bond polarities, from where the hydrogen is polarized δ+ (*e.g.* H–S) to where it is δ– (*e.g.* B–H or Ga–H). The most important covalent hydrides formed by the p-block elements are summarized in Table 2.2. Elements in Groups 14 to 17 form hydrides with 'normal' covalent bonds.

Some elements which form covalent hydrides, and their electronegativities (Pauling scale):
P   2.2
S   2.5

MgH$_2$ is intermediate between covalent and ionic. CuH, ZnH$_2$ and CdH$_2$ are intermediate between covalent and metallic.

Table 2.2 Important covalent hydrides of the p-block elements

Group	13	14	15	16	17
	B$_2$H$_6$	C$_n$H$_{2n+2}$ᵃ	NH$_3$	H$_2$O	HF
		C$_n$H$_{2n}$	N$_2$H$_4$	H$_2$O$_2$	
		C$_n$H$_{2n-2}$ *etc.*			
	(AlH$_3$)$_n$	Si$_n$H$_{2n+2}$ ($n \leq 8$)	PH$_3$	H$_2$S	HCl
			P$_2$H$_4$	H$_2$S$_n$	
		Ge$_n$H$_{2n+2}$ ($n \leq 9$)	AsH$_3$	H$_2$Se	HBr
		SnH$_4$	SbH$_3$	H$_2$Te	HI

ᵃThere is no apparent limit to the length of chains formed by carbon atoms.

C$_n$H$_{2n+2}$ = alkanes
C$_n$H$_{2n}$ = alkenes
C$_n$H$_{2n-2}$ = alkynes

Worked Problem 2.4

Q Using electronegativities (Ga 1.81, Bi 2.02, Pb 2.33), discuss whether the hydrides Ga$_2$H$_6$, BiH$_3$ and PbH$_4$ are expected to be covalent.

A Ga, Bi and Pb, although metals, have electronegativities greater than 1.5, so the hydrides are substantially covalent in character.

A second major class of hydrides are the electron-deficient hydrides, typified by B$_2$H$_6$ (and higher hydrides of boron) (Section 5.6).

Worked Problem 2.5

Q Explain why the enthalpies of formation, $\Delta_f H°$ (kJ mol^{-1}) of the hydrogen halides become more positive in the order HF (–273) > HCl (–92) > HBr (–36) > HI (+26.5).

A Enthalpy of formation refers to the reaction $0.5H_2 + 0.5X_2 \rightarrow HX$. For HF, the breaking of a strong H–H bond (and a weak F–F bond) is easily compensated by formation of a very strong H–F bond. Going down the group, the X–X bonds become weaker, but the formation of an increasingly weak H–X bond (owing to poor overlap of a small hydrogen 1s orbital with the large diffuse halogen orbital) cannot compensate for the strength of the H–H bond which must be broken. The $\Delta_f H°$ values therefore become more positive going down Group 17.

Synthesis

Covalent hydrides can be synthesized by several methods:

- Direct combination of the elements (equations 2.12 and 2.13).

$$2H_2 + O_2 \rightarrow 2H_2O \qquad (2.12)$$

$$H_2 + Cl_2 \rightarrow 2HCl \qquad (2.13)$$

- Reduction of a halide or oxide (*e.g.* equation 2.14); this is generally the most widely applicable method.

$$SiCl_4 + LiAlH_4 \rightarrow SiH_4 + LiAlCl_4 \qquad (2.14)$$

- Hydrolysis of a metal phosphide, carbide, silicide, boride, *etc.* (*e.g.* equation 2.15).

$$Ca_3P_2 + 6H_2O \rightarrow 2PH_3 + 3Ca(OH)_2 \qquad (2.15)$$

- Interconversion of hydrides, for example with an electrical discharge (*e.g.* equation 2.16).

$$nGeH_4 \rightarrow Ge_2H_6 + Ge_3H_8 + \text{higher hydrides} \qquad (2.16)$$

A reducing agent is easily oxidized; this is the process operating in equations 2.17 and 2.18

Generally, hydrides of p-block elements are reducing agents; some (such as SiH_4) spontaneously combust in air (*e.g* equation 2.17), while others such as CH_4 require a spark to initiate the reaction (equation 2.18):

$$SiH_4 + 8O_2 \rightarrow SiO_2 + 2H_2O \qquad (2.17)$$

$$CH_4 + 2O_2 \rightarrow CO_2 + 2H_2O \qquad (2.18)$$

Worked Problem 2.6

Q Balance the following reactions, which all produce hydrides:
(a) $Na_2S + HCl \rightarrow$
(b) $PhPCl_2 + LiAlH_4 \rightarrow$

A (a) $Na_2S + HCl \rightarrow H_2S + 2NaCl$ (H_2S is a weaker acid than HCl).
(b) $2PhPCl_2 + LiAlH_4 \rightarrow 2PhPH_2 + LiAlCl_4$ ($PhPCl_2$ is an organic derivative of PCl_3).

Worked Problem 2.7

Q How would you prepare the following hydrides: (a) HF and (b) SeH_2?

A (a) $H_2SO_{4(l)} + NaF_{(s)} \rightarrow HF_{(g)} + NaHSO_{4(s)}$. Alternatively, the (explosive) reaction between H_2 and F_2 could be used.
(b) H_2 does not react with elemental Se, so an indirect method must be used:

$$2Na + Se \rightarrow Na_2Se \text{ (or } Se + NaBH_4 \rightarrow NaHSe), \text{ then}$$
$$Na_2Se + HCl \rightarrow H_2Se + 2NaCl$$
$$\text{(or } NaHSe + HCl \rightarrow H_2Se + NaCl)$$

Both Na and Na[BH$_4$] are reducing agents.

2.4.3 Metallic (Interstitial) Hydrides

A number of the transition metals, lanthanides and actinides absorb hydrogen to variable extents, giving metallic hydrides which have many of the properties of the metals, such as hardness, conductivity and lustre. These hydrides are non-stoichiometric, with compositions such as $PdH_{0.6}$ and $VH_{1.6}$, with the hydrogens in tetrahedral sites ('interstices') in a close-packed metal lattice. The lanthanide hydrides can also take up hydrogens in octahedral sites, giving *ionic* MH_3 phases.

2.5 Compounds containing Hydrogen formally as H⁺

Hydrogen is able to form compounds by loss of its sole 1s electron, giving the H^+ ion. The conversion of H_2 into *gas-phase* protons requires a large amount of energy (equation 2.19).

$$H_{2(g)} - 2e^- \rightarrow 2H^+_{(g)} \quad \Delta H = +\ 3054 \text{ kJ mol}^{-1} \qquad (2.19)$$

If a proton is generated in aqueous solution, it is solvated by water to form the **hydroxonium ion**, H_3O^+, sometimes called **hydronium** or **oxonium**. This ion has hydrogens bearing a partial positive charge, and so it will be further solvated by water molecules of the bulk solvent. A proton dissolved in water can be approximated by the species $[H_3O(H_2O)_3]^+$, or $[H_9O_4]^+$, which has the structure shown in Figure 2.1, where each H atom of H_3O^+ is hydrogen bonded (Section 2.6) to the O atom of a water molecule.[1]

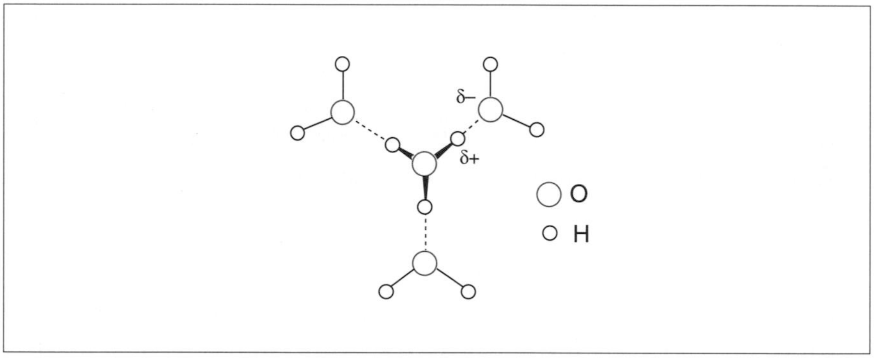

Figure 2.1 The structure of $H_9O_4^+$

2.6   The Hydrogen Bond

2.6.1   General Features

When hydrogen is bonded to the highly electronegative elements F, O, Cl or N, a secondary interaction – a **hydrogen bond** – occurs between the partially positively charged hydrogen of one molecule and a partially negatively charged atom.

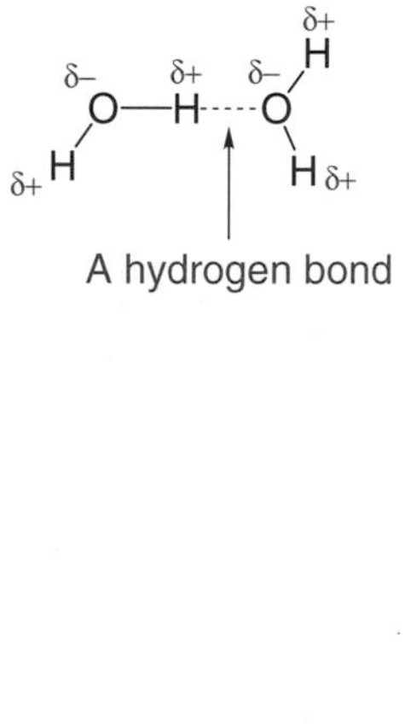

This hydrogen bonding effect is only important for hydrogen (and not other electropositive elements) because of the small size of hydrogen, and the absence of inner electron shells which could shield the nucleus.

Hydrogen bonding is clearly illustrated by comparing the boiling points of the hydrides of the Group 16 elements (EH_2) with the Group 14 elements (EH_4), shown in Figure 2.2. For the EH_4 compounds there is an increase in boiling point with relative molar mass (owing to increasing van der Waals forces), and the same trend is observed for H_2S, H_2Se and H_2Te. However, the boiling point of water is anomalously high, and this is due to extensive hydrogen bonding (Section 2.6.2). Hydrogen bonding does not occur to any significant extent in H_2S or the other heavier hydrides. The same trend for the Group 16 dihydrides is also seen on examination of:

A hydrogen bond between two water molecules:

A hydrogen bond

3

The Group 1 (Alkali Metal) Elements: Lithium, Sodium, Potassium, Rubidium, Caesium and Francium

Aims

By the end of this chapter you should understand:

- The dominance of the M^+ cations in the overall chemistry of this group
- The high reactivities (reducing properties) of the metals
- The formation of basic oxides and hydroxides
- The similarities in the chemistries of lithium and magnesium

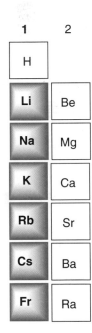

3.1 Introduction and Oxidation State Survey

The chemistry of the alkali metals is dominated by the tendency to lose the single s-electron and attain a noble gas electronic configuration. The elements have low first and very high second ionization potentials, and so the dominant feature of their chemistry is the formation of M^+ cations. Going down the group, there is a decrease in first ionization energy, as the valence ns electron is further from the nucleus (Figure 3.1). For Rb and Cs the values are higher than expected, because these elements have filled inner d-shell(s) with relatively poor shielding of the outer ns electron. There have been some indications that caesium may be able to exist in a higher oxidation state (+3), though no compounds have yet been isolated.[1]

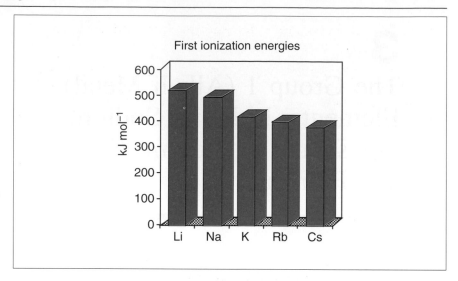

Figure 3.1 The first ionization energies of the Group 1 metals

In Group 1, there is less variation in the chemistry than for elements of any other main group, though there are some subtle trends in the degree of solvation and complex formation (which is most important for the small Li^+ ion) and in the stabilities of certain compounds, such as carbonates and nitrates (Section 3.4).

3.2   The Elements

Sodium is the most abundant alkali metal, occurring in large underground deposits of sodium chloride. Seawater contains a high concentration (10,800 ppm) of Na^+, and 390 ppm of K^+. All isotopes of francium are radioactive. It is formed naturally in trace amounts by the decay of actinium.

The chlorides, MCl, are the common commercial products; lithium carbonate is also made on a large scale. Lithium and sodium metals are obtained by electrolysis of molten LiCl or NaCl (equation 3.1).

$$NaCl \rightarrow Na + 0.5Cl_2 \qquad (3.1)$$

Potassium is obtained by reduction of KCl with sodium vapour at 850 °C (equation 3.2).

$$KCl + Na \rightleftharpoons NaCl + K \qquad (3.2)$$

The alkali metals crystallize in body-centred cubic lattices (Figure 3.2).

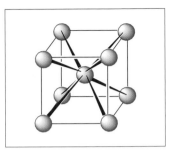

Figure 3.2 Body-centred cubic lattice adopted by the alkali metals

---

Worked Problem 3.1

Q In the gas phase the alkali metals form dimers M_2. Draw a simple molecular orbital (MO) diagram for an M_2 diatomic molecule.

A Like hydrogen, the alkali metals have an ns orbital available, containing one electron. Overlap of these gives bonding and antibonding σ MOs:

Molecular orbital diagram of M_2

Each M atom provides one electron, filling the σ bonding MO, but leaving the antibonding MO σ* empty, giving a net M–M bond order of 1.

See Section 1.5 for discussion of MO theory.

3.3 Chemistry of the Alkali Metals

All the alkali metals are highly reactive, combining readily with most other elements, often explosively. Almost all derivatives are ionic, though some lithium compounds (especially the organolithium compounds) can have appreciable covalent character. The metals themselves are very powerful reducing agents. Li has the most negative $E°$ value (see Table 3.1), because of the very high hydration energy (Box 3.1) of the Group 1 metal cations. However, in terms of absolute reactivity the metals become much more reactive going down the group.

The formation of monovalent M^+ cations dominates the chemistry of the Group 1 metals.

Table 3.1 Standard redox potentials for the alkali metals

Equation	$E°$ value (V)
$Li^+ + e^- \rightleftharpoons Li$	−3.05
$Na^+ + e^- \rightleftharpoons Na$	−2.71
$K^+ + e^- \rightleftharpoons K$	−2.94
$Rb^+ + e^- \rightleftharpoons Rb$	−2.94
$Cs^+ + e^- \rightleftharpoons Cs$	−3.02

3.4   Simple Salts of the Alkali Metals

Some insoluble potassium, rubidium and caesium salts:

Salt	Anion
$M[ClO_4]$	Chlorate(VII) (perchlorate)
$M_2[PtCl_6]$	Hexachloroplatinate(IV)
$M_2[SiF_6]$	Hexafluorosilicate
$M[BPh_4]$	Tetraphenylborate

Salts of the alkali metals tend to be very soluble in water, especially for lithium and sodium, though a few salts of potassium, rubidium and caesium are also insoluble, and LiF and Li_2CO_3 are also only sparingly soluble in water. The solubility of a salt is dependent on two opposing, and typically large, energies: the **lattice energy** (holding the solid together) and the **hydration energy** of the ions (which tends to dissolve the solid).

Box 3.1 Lattice Energy and Hydration Energy

The **lattice energy** (U) for a salt MX refers to the process:

$$M^+_{(g)} + X^-_{(g)} \rightarrow MX_{(s)}$$

and is proportional to z^+z^-/r, where z^+ and z^- = charge on cation and anion, respectively, and r = sum of ionic radii of cation and anion. A high lattice energy of an ionic solid results in a high melting point.

Hydration energy refers to the process:

$$M^+_{(g)} + nH_2O_{(l)} \rightarrow [M(H_2O)_n]^+_{(aq)}$$

Lattice and hydration energies are generally at a maximum for a small cation–small anion combination; LiF is the least soluble alkali metal fluoride because of its very high lattice energy. For the few insoluble heavier alkali metal salts, the lattice energy is presumably larger than the relatively low hydration energy for the larger and more poorly solvated M^+ cations.

Worked Problem 3.2

Q Using the thermochemical data below, calculate the standard enthalpy of formation ($\Delta_f H°$) of $NaCl_2$, and comment on the factors which make it an unstable compound.

Process	$\Delta H°$ (kJ mol^{-1})
$Na_{(s)} \rightarrow Na_{(g)}$	+107
$Na_{(g)} \rightarrow Na^+_{(g)} + e^-$	+502
$Na^+_{(g)} \rightarrow Na^{2+}_{(g)} + e^-$	+4569
$Cl_{2(g)} \rightarrow 2Cl_{(g)}$	+242
$Cl_{(g)} \rightarrow Cl^-_{(g)}$	−360

Calculated lattice energies for $NaCl$ and $NaCl_2$ are -760 and -2250 kJ mol^{-1}, respectively.

A Construct a thermodynamic (Born–Haber) cycle, such that the only unknown is $\Delta_f H°$.

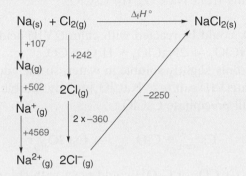

Thus $\Delta_f H° = 107 + 502 + 4569 + 242 - 720 - 2250 = +2450$ kJ mol^{-1}. With a large, positive $\Delta_f H°$, the compound is clearly unstable. The principal factor for the exclusive formation of M^+ ions for the alkali metals is the extremely high second ionization energy of sodium (4569 kJ mol^{-1}), which cannot be compensated for by the larger lattice energy of $NaCl_2$ compared to $NaCl$.

The metals react with water, lithium relatively slowly, but with increasing vigour going down the group, giving the hydroxide plus hydrogen gas (equation 3.3).

$$2M + 2H_2O \rightarrow 2MOH + H_2 \qquad (3.3)$$

The hydroxides are useful starting precursors for the synthesis of other salts by neutralization with the appropriate acid. The hydroxides absorb CO_2 to give the alkali metal carbonate M_2CO_3 or (with all except lithium) the hydrogencarbonate (bicarbonate) $MHCO_3$. The hydrogencarbonates decompose on heating to give the carbonates, which themselves decompose on stronger heating to give the oxide plus carbon dioxide (equation 3.4).

$$2MHCO_3 \rightarrow M_2CO_3 \; (+ H_2O + CO_2) \rightarrow M_2O \; (+ CO_2) \qquad (3.4)$$

The nitrates MNO_3 (M = Na, K, Rb, Cs) decompose to the **nitrites** on heating (equation 3.5), while, in contrast, $LiNO_3$ decomposes to Li_2O.

The small, highly polarizing Li^+ ion aids decomposition of larger oxyanions, *e.g.* NO_3^- and CO_3^{2-}; the Li_2O that is formed has a high lattice energy, which is a driving force for the decomposition.

$$2MNO_3 \rightarrow 2MNO_2 + O_2 \qquad\qquad (3.5)$$

Worked Problem 3.3

Q How would you synthesize the following compounds: (a) $NaClO_4$ starting from Na_2CO_3; (b) $CsClO_4$.

A (a) Na_2CO_3 could be reacted with chloric(VII) acid, $HClO_4$:
$Na_2CO_3 + 2HClO_4 \rightarrow 2NaClO_4 + H_2O + CO_2$
(b) $CsClO_4$ is only slightly soluble in water, so the addition of any soluble chlorate(VII) salt (*e.g.* $NaClO_4$) to any soluble caesium salt (*e.g.* $CsCl$) will precipitate $CsClO_4$:

$$Cs^+_{(aq)} + ClO_4^-{}_{(aq)} \rightarrow CsClO_{4(s)}$$

Alternatively, Cs_2CO_3 or $CsOH$ could react with chloric(VII) acid, analogous to (a).

The **alkali metal halides** are well-known compounds. CsCl, CsBr and CsI crystallize with the same structure (the caesium chloride type) (Figure 3.3), while the other salts crystallize in the sodium chloride structure (Figure 3.4). In the CsCl structure the cation and anion are both eight coordinate, while in the NaCl structure they are six coordinate. Both the CsCl and NaCl structure types are important, and commonly adopted by many other ionic compounds with a 1:1 ratio of cation and anion.

In the CsCl structure, each Cl ion only contributes 1/8th its volume to the unit cell shown in Figure 3.3. Thus there is one cation per anion.

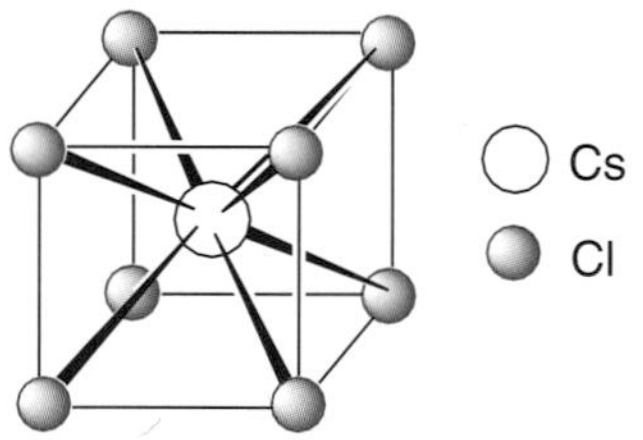

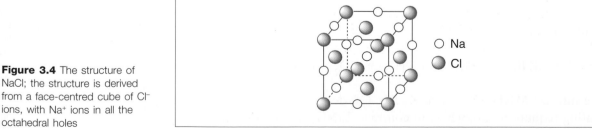

Figure 3.3 The structure of CsCl; each of the Cl⁻ ions is also surrounded by a cube of Cs⁺ ions from adjacent unit cells

Figure 3.4 The structure of NaCl; the structure is derived from a face-centred cube of Cl⁻ ions, with Na⁺ ions in all the octahedral holes

Box 3.2 Predicting Structures using Radius Ratio Rules

The structure adopted by a salt M^+X^- can be predicted by considering the relative sizes of the cation (radius r_+) and anion (radius r_-), the so-called radius ratio rules:

r_+/r_-	Structure type
0.225–0.414	ZnS (sphalerite); see Section 11.5 and Figure 11.1
0.414–0.732	NaCl (Figure 3.4)
>0.732	CsCl (Figure 3.3)

Radius ratio rules provide a reasonably general means of assessing the likely structure adopted by an *ionic* solid, but give incorrect predictions when:

- there is significant covalent bonding
- the radius ratio is near a borderline
- ionic radii are not known accurately; values vary with the coordination number (how many nearest neighbour ions) of the ion

Worked Problem 3.4

Q Assuming the following ionic radii (in pm) [Li^+ 74; Cs^+ 167; F^- 133; I^- 220] predict what structure type will be adopted by the following alkali metal salts: (a) CsI; (b) LiF.

A (a) CsI. The radius ratio r_+/r_- is 167/220 = 0.76, so the CsCl structure is predicted.
(b) LiF. The radius ratio r_+/r_- is 74/133 = 0.56, so the NaCl structure is predicted.

3.5 Compounds with Oxygen and Sulfur

The products of combustion of the alkali metals in air or oxygen vary with the metal; three different oxygen-containing anions are formed, which are distinguished by the different products formed by hydrolysis with water (Table 3.2).

The oxides of the Group 1 metals are *basic*, since they react with water to give the hydroxide, *e.g.* $Na_2O + H_2O \rightarrow 2NaOH$

Table 3.2 Products of combustion of alkali metals in air, and the products formed upon hydrolysis of these oxides

Metal	Salt	Anion	Hydrolysis product(s) of salt
Li	Li_2O	O^{2-} (oxide)	OH^-
Na	Na_2O_2 + Na_2O	O_2^{2-} (peroxide)	OH^-, H_2O_2
K, Rb, Cs	MO_2	O_2^- (superoxide)	OH^-, H_2O_2, O_2

The bonding in the superoxide and peroxide anions is described in Chapter 1, and Chapter 8 discusses peroxides in more detail.

Lithium does not form a superoxide, and lithium peroxide (Li_2O_2, formed by the reaction of LiOH with hydrogen peroxide, H_2O_2) is unstable, decomposing to Li_2O and oxygen gas. Going down the group, the superoxides and peroxides become more stable; this is another example of the stabilization of a large anion by a large cation (compare nitrates and carbonates, Section 3.4). The monoxides (M_2O) of sodium to caesium can be formed from oxygen and an excess of the metal (to avoid the formation of the superoxide and/or peroxide). The excess of metal is removed by evaporation. An improved synthesis is to reduce the metal nitrite with metal (equation 3.6).

$$6M + 2MNO_2 \rightarrow 4M_2O + N_2 \qquad (3.6)$$

Ozone: see Section 8.2.1.

Reaction of the alkali metals with ozone (O_3) gives ozonide salts (MO_3), which are paramagnetic. The alkali metals react with sulfur to form sulfides and polysulfides of composition M_2S_x; these are discussed separately in Section 8.3.5.

3.6   Compounds with Nitrogen

The nitride anion N^{3-} is isoelectronic with carbide (C^{4-}), oxide (O^{2-}) and fluoride (F^-).

Lithium is the only alkali metal which combines with nitrogen gas, giving lithium nitride (equation 3.7). The salt is presumably stable because of the very high lattice energy from the very small Li^+ ion and the small, highly charged N^{3-} ion, despite having to break the strong N≡N bond. Hydrolysis of Li_3N gives ammonia (equation 3.8).

$$6Li + N_2 \rightarrow 2Li_3N \qquad (3.7)$$

$$Li_3N + 3H_2O \rightarrow 3LiOH + NH_3 \qquad (3.8)$$

When alkali metals are dissolved in anhydrous liquid ammonia, intensely blue, electrically conducting solutions are formed. These solutions contain solvated ('ammoniated') alkali metal cations together with solvated electrons, and are powerful reducing agents. Addition of a catalytic amount of a transition metal complex [such as iron(III) nitrate] results in a reaction to form the **alkali metal amide** (MNH_2) and hydrogen gas (equation 3.9).

$$2M + 2NH_3 \rightarrow 2M^+NH_2^- + H_2 \qquad (3.9)$$
$$(\text{or } 2e^- + 2NH_3 \rightarrow 2NH_2^- + H_2)$$

Liquid ammonia is used as a **non-aqueous solvent**, and there are many comparisons with water and related compounds. Just as sodium hydroxide is a base in water, sodium amide ($NaNH_2$) is a base in liquid ammonia, because it is able to deprotonate acidic molecules R–H (equation 3.10).

See also Worked Problem 7.1.

$$R–H + NaNH_2 \rightarrow NaR + NH_3 \qquad (3.10)$$

3.7 Hydrides

The alkali metals combine with hydrogen on heating to give ionic ('salt-like') hydrides, containing the colourless hydride ion, H^- (equation 3.11).

$$2M + H_2 \rightarrow 2M^+H^- \qquad (3.11)$$

The hydrides have the sodium chloride structure (Figure 3.4), and readily hydrolyse on reaction with water (equation 3.12). The H^- anion is a strong base, and can be used to deprotonate organic molecules containing relatively acidic C–H groups, *e.g.* equation 3.13.

Further details on ionic hydrides are given in Section 2.4.1.

Other common hydride materials which contain alkali metal cations are sodium borohydride, $Na[BH_4]$, and lithium aluminium hydride, $Li[AlH_4]$, which are discussed in Chapter 5.

$$Na^+H^- + H_2O \rightarrow NaOH + H_2 \qquad (3.12)$$

$$Na^+H^- + CH_3S(O)CH_3 \rightarrow Na^+[CH_3S(O)CH_2]^- + H_2 \qquad (3.13)$$

3.8 Compounds with Carbon

The alkali metals readily react with ethyne (acetylene, HC≡CH) to form metal **ethynides** (acetylides), containing singly or doubly deprotonated ethyne (equation 3.14).

$$2HC \equiv CH \xrightarrow{\ M\ } 2M(C \equiv CH)\ (+H_2) \xrightarrow{\ M\ } 2M_2(C \equiv C)\ (+H_2) \quad (3.14)$$

In this reaction the ethyne is reacting as an acid.

There is an extensive **organometallic chemistry** of the alkali metals, especially for lithium. Lithium metal reacts with alkyl or aryl chlorides (and bromides) in a dry hydrocarbon or ether solvent to give **organolithium** compounds (equation 3.15), which are largely covalent.

An organometallic compound is one which formally contains a metal–carbon bond (except to cyanide).

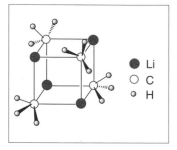

● Li
○ C
∘ H

Figure 3.5 The tetrameric structure of methyllithium, $(CH_3Li)_4$

$[Li(NH_3)_4]^+I^-$ is an example of a simple coordination complex:

$$NH_3$$

$$Li^+$$

$$H_3N \quad NH_3$$
$$NH_3$$

In 18-crown-6 there are 18 atoms in the ring, of which six are oxygens. In cryptand [2,2,2] there are two oxygen atoms in each of the three 'branches' linking the two nitrogens together.

$$R{-}X + 2Li \rightarrow LiR + LiX \qquad (3.15)$$

Organolithium compounds are usually extensively aggregated together in the solid and solution states; for example, methyllithium exists as a tetramer, $CH_3Li)_4$, which has the structure shown in Figure 3.5.

The organolithium compounds have similarities to the organomagnesium compounds RMgX and R_2Mg (R = alkyl or aryl) (see Section 4.7).

3.9   Alkali Metal Complexes

Because of their relatively low **charge density**, the alkali metals form relatively few complexes with neutral ligands, though lithium, with the highest charge density, forms the most stable complexes. Lithium salts are often more soluble than their sodium counterparts in solvents such as alcohols and ethers owing to the coordination of the solvent oxygen atoms to the lithium cation. Lithium ions are generally four coordinate, while sodium and potassium ions are six coordinate.

The most stable complexes are formed by **multidentate ligands**, which can encapsulate the metal ion, forming numerous metal–ligand bonds. Examples include **crown ethers** and **cryptands**; an example of each class of ligand is shown in Figure 3.6.

When the size of the cavity in the ligand is changed (by varying the number of $-OCH_2CH_2-$ groups), the selectivity for different alkali metal cations with different ionic radii can be altered.

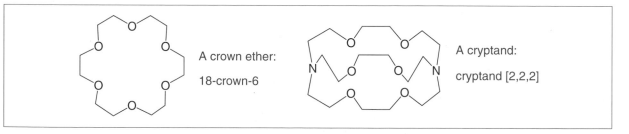

A crown ether:

18-crown-6

A cryptand:

cryptand [2,2,2]

Figure 3.6 Examples of some multidentate ligands which form stable alkali metal complexes

---

Worked Problem 3.5

Q Figure 3.7 shows the equilibrium constants ('stability constants') for
$M^+ + cryptand \rightleftharpoons [M(cryptand)]^+$, where cryptand = cryptand [2,2,2]. Comment on the form of the graph.

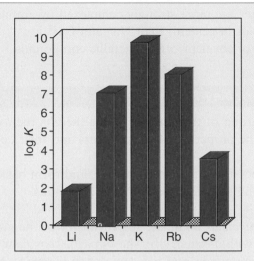

Figure 3.7 Plot of log$K$ (K = stability constant) for complexes of cryptand [2,2,2] with alkali metal cations

A The potassium ion fits the cavity size in cryptand [2,2,2] very well, so the complex is stable and has a large stability constant. The smaller Li$^+$ and Na$^+$ ions are too small to make good bonding contacts with the ligand atoms, and the cations 'rattle around' inside the cavity. The complexes are less stable than those of potassium. For rubidium and caesium ions, these are larger than K$^+$, and the cryptand ligand must distort in order to form complexes; again the complexes are less stable (with smaller log K values) than for potassium.

3.10 Similarities in the Chemistry of Lithium and Magnesium

Some of the properties of the main group elements can be rationalized in part by the **diagonal relationship**, where an element has a similarity to the element 'one down to the right'. Thus, lithium shows many similarities to magnesium rather than to the heavier alkali metals sodium and potassium. The higher charge of the Mg^{2+} ion is partly offset by its larger size, so Mg^{2+} and Li$^+$ have similar **charge densities**.

Some of the similarities between lithium and magnesium are:

- Direct formation of the nitride and carbide
- Combust in air to the normal oxides, Li$_2$O and MgO (sodium forms mostly the peroxide Na$_2$O$_2$)
- Stability of oxysalts, *e.g.* Li$_2$CO$_3$ and MgCO$_3$ give Li$_2$O and MgO on heating; Na$_2$CO$_3$ is stable to moderate heating.

- Li^+ and Mg^{2+} are more strongly complexed by ammonia (and other donor ligands in general) than Na^+ and Ca^{2+} ions
- Formation of covalent organometallic compounds

Summary of Key Points

1. In this group of the Periodic Table there is the *least variation* among the properties of the elements and their compounds.

2. Alkali metals are all *highly reactive*, and form basic oxides and water-soluble halides, containing M^+ cations.

3. *Ionic compounds* dominate the chemistry of the group, though lithium, with its small size, shows partial covalent character in many compounds, and has some resemblance to magnesium.

Problems

3.1. What are the oxidation states of the alkali metals in the following substances: (a) Na metal; (b) $NaPF_6$; (c) RbO_2; (d) Na_2(18-crown-6).

3.2. Identify the element X in each of the following:
(a) The reaction between metallic X and water is quite slow.
(b) X is a reactive metal which forms three ionic oxides with compositions X_2O, XO_2 and X_2O_2; the perchlorate $XClO_4$ is insoluble in water.
(c) X is the most abundant element in Group 1.
(d) The chemistry of X resembles that of magnesium.

3.3. Explain why, on going from Li to Cs: (a) the first ionization energy of the metal decreases; (b) the hydration energy of the M^+ cation decreases.

3.4. How would you prepare the following compounds, starting from lithium metal: (a) Li_2CO_3; (b) PhLi; (c) $LiNH_2$.

3.5. Which of the following statements is incorrect:
(a) The halides of all the Group 1 metals are largely ionic.
(b) RbO_2 is stable because the large Rb^+ ion stabilizes the anion.
(c) All Group 1 oxides are basic.

(d) The oxidation state of potassium in K_2O_2 is +2.

3.6. Explain why the melting points of the alkali metal bromides decrease in the order: NaBr (747 °C) > KBr (734 °C) > RbBr (693 °C) > CsBr (636 °C).

3.7. Lithium salts of oxyanions tend to be less stable than the analogous sodium or potassium salts. Give three examples of lithium salts which decompose more readily.

3.8. Assuming the following ionic radii (pm) [Li^+ 74; Na^+ 102; Rb^+ 149; I^- 220; BF_4^- 218], using radius ratio rules (Box 3.2) predict what solid-state structure will be adopted by: (a) RbI; (b) NaI; (c) $LiBF_4$.

Reference

1. K. Moock and K. Seppelt, *Angew. Chem., Int. Ed. Engl.*, 1989, **28**, 1676.

Further Reading

Alkali and alkaline earth metal cryptates, D. Parker, *Adv. Inorg. Chem. Radiochem.*, 1983, **27**, 1.
Electrides, negatively charged metal ions and related phenomena, J. L. Dye, *Prog. Inorg. Chem.*, 1984, **32**, 327.

4

The Group 2 Elements: Beryllium, Magnesium, Calcium, Strontium, Barium and Radium

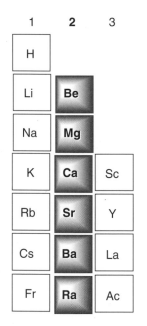

The elements calcium, strontium and barium are often referred to as the **alkaline earth** metals

Aims

By the end of this chapter you should understand:

- The differences between beryllium and the other members of the group, and between the Group 1 and Group 2 elements
- The dominance of the +2 oxidation state
- The formation of coordination complexes

4.1 Introduction and Oxidation State Survey

The Group 2 metals are chemically similar in a number of respects to the Group 1 metals. All are true metals, but there are some important differences between the two groups, and within the Group 2 metals themselves. All Group 2 metals form M^{2+} ions, but the high charge density of Be^{2+} means that the free ion cannot exist, and all its compounds are either covalent or contain solvated ions such as $[Be(H_2O)_4]^{2+}$. Beryllium therefore shows properties markedly different from the other elements in Group 2. The higher charge density of the 2+ ions for the Group 2 metals, when compared to the 1+ charge of the Group 1 metals, results in the M^{2+} ions being markedly smaller. They are therefore more strongly hydrated and form more complexes than the alkali metals. In addition, because of the higher lattice energies (see Box 3.1) involving 2+ ions, many salts are less soluble than their Group 1 metal analogues. For example, potassium sulfate is readily soluble in water, but calcium and strontium sulfates are not.

The diagonal relationship in the Periodic Table applies well for these main group elements; the similarities between Li and Mg have been mentioned in Section 3.11, and those between Be and Al are discusssed in Section 4.10.

4.2   The Elements

Beryllium is a relatively rare element, but occurs mainly in the mineral *beryl*, $Be_3Al_2Si_6O_{18}$ (see Section 12.3), which is heated with $Na_2[SiF_6]$ to give $Na_2[BeF_4]$. Sodium hydroxide is then added to give beryllium hydroxide (equation 4.1), which reacts with $NH_4[HF_2]$ to give $(NH_4)_2[BeF_4]$ and then, on heating, BeF_2 (equation 4.2). Beryllium metal is obtained by magnesium reduction of BeF_2 (equation 4.3).

$$Na_2BeF_4 + 2NaOH \rightarrow Be(OH)_2 + 4NaF \quad (4.1)$$

$$Be(OH)_2 + 2NH_4[HF_2] \rightarrow (NH_4)_2[BeF_4] \xrightarrow{heat} BeF_2 \quad (4.2)$$

$$BeF_2 + Mg \rightarrow MgF_2 + Be \quad (4.3)$$

Emerald is the green gem form of beryl.

Beryllium and its compounds are extremely toxic, second only to plutonium and its compounds. As a result, chemists often prefer to undertake theoretical rather than practical beryllium chemistry.

Magnesium is widespread in the Earth's crust, but the main commercial sources are from seawater and the minerals *magnesite*, $MgCO_3$, and *dolomite*, $(Ca,Mg)CO_3$. **Calcium** similarly occurs widely as limestone, $CaCO_3$. **Strontium** and **barium** occur naturally in several minerals including *celestine*, $SrSO_4$, and *barytes*, $BaSO_4$. All isotopes of radium are radioactive; it occurs naturally as a decay product of uranium, and is long-lived enough to be isolated as a pure substance.

4.3   Simple Compounds and Salts

The simple *anhydrous* compounds of beryllium tend to be covalent in nature, though when crystallized from water, salts containing the hydrated $[Be(H_2O)_4]^{2+}$ ion can be formed. The $[Be(H_2O)_4]^{2+}$ ion, similar to $[Al(H_2O)_6]^{3+}$, is acidic, as a result of the high polarizing power of the small Be^{2+} ion which results in hydrolysis (equation 4.4). The other hydrated Group 2 cations are not acidic, owing to their lower charge densities.

Zinc and beryllium show some common chemical properties; see Section 11.1.

$$[Be(H_2O)_4]^{2+} + H_2O \rightleftharpoons [Be(OH)(H_2O)_3]^+ + H_3O^+ \quad (4.4)$$

$[Be(H_2O)_4]^{2+}$ therefore only exists in strong acid solutions. On increasing the pH, hydroxide-bridged ions (containing Be–OH–Be groups) such as $[Be(OH)_3]_3^{3-}$ form, before the precipitation of $Be(OH)_2$. In excess hydroxide, BeO and $Be(OH)_2$ dissolve to give the beryllate ion, $[Be(OH)_4]^{2-}$, demonstrating the **amphoteric** nature of beryllium.

Beryllium chloride ($BeCl_2$) is a covalent polymer in the solid state,

$[Be(OH)_3]_3^{3-}$

An amphoteric oxide is one which reacts with both acid and base; Al_2O_3 is another example.

containing Be–Cl–Be bridges (Figure 4.1). In the vapour the polymer is broken up to give a mixture of the monomer ($BeCl_2$) and the dimer (Be_2Cl_4), which are in equilibrium (Figure 4.1). Anhydrous beryllium halides are soluble in many solvents, owing to the formation of complexes.

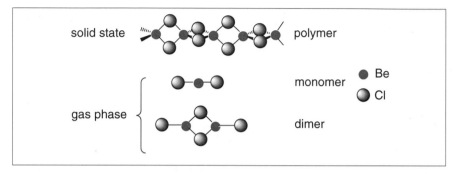

Figure 4.1 Structures of $BeCl_2$ in the solid state and gas phase

The **chlorides**, **bromides** and **iodides** of Mg, Ca, Sr and Ba are all typical ionic, water-soluble salts, but the fluorides are only slightly water soluble, owing to the high *lattice energies* (see Box 3.1) for dipositive cations and small F^- ions. The structure of *fluorite*, CaF_2 (Figure 4.2), is adopted by many MX_2 ionic solids. The structure is based on a face-centred cubic lattice of Ca^{2+} ions, with F^- ions in all of the tetrahedral holes (two per Ca).

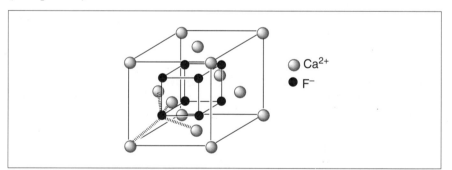

Figure 4.2 The structure of CaF_2. The tetrahedral coordination around F^- is indicated by dashed lines

Worked Problem 4.1

Q Calculate the lattice energy (U) of CaF_2 using the thermochemical data below, and compare the value with the lattice energy of SrF_2 (-2496 kJ mol^{-1}).

Process	$\Delta H°$ (kJ mol^{-1})
$Ca_{(s)} \rightarrow Ca_{(g)}$	+178
$Ca_{(g)} \rightarrow Ca^{2+}_{(g)}$	+1748
$F_{2(g)} \rightarrow 2F_{(g)}$	+158

$$F_{(g)} + e^- \rightarrow F^-_{(g)} \qquad\qquad -334$$
$$Ca_{(s)} + F_{2(g)} \rightarrow CaF_{2(s)} \qquad -1220$$

A Construct a thermodynamic (Born–Haber) cycle, with the lattice energy as the unknown:

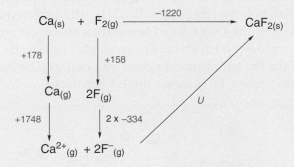

Hence, $-1220 = 178 + 1748 + 158 - 668 + U$

Therefore $U = -2636$ kJ mol^{-1}. This is larger than the lattice energy of SrF$_2$ since the Ca^{2+} ion is smaller than the Sr^{2+} ion; see Box 3.1.

Compare Worked Problem 3.2.

The **sulfates** MSO$_4$ are all known, and decrease in solubility going from the soluble BeSO$_4$ to sparingly soluble BaSO$_4$. All the Group 2 metal sulfates lose SO$_3$ to form the oxide on strong heating (equation 4.5).

$$MSO_4 \rightarrow MO + SO_3 \qquad\qquad (4.5)$$

The insolubility of BaSO$_4$ is used as a qualitative test for sulfate (or barium) ions, and its opacity to X-rays allows it to be used as a non-toxic imaging agent for diagnosing gastrointestinal conditions, such as ulcers.

Worked Problem 4.2

Q Explain why the decomposition temperature for BeSO$_4$ is relatively low (580 °C) compared with SrSO$_4$ (1374 °C)

A The Be^{2+} ion has a high charge density and is strongly polarizing. The resulting decomposition product, BeO, has a high lattice energy because it contains small and doubly charged ions (see Box 3.1). In contrast, the large Sr^{2+} ion is much less polarizing, and SrO has a lower lattice energy than BeO.

The tendency for beryllium salts to decompose more readily than their other Group 2 counterparts is again illustrated by the carbonates MCO$_3$, which increase in stability going down the group. Beryllium carbonate only exists when crystallized under an atmosphere of CO$_2$, while magnesium, calcium and strontium carbonates give the oxide only on heating (equation 4.6). Barium carbonate is stable to heat.

$$MgCO_3 \rightarrow MgO + CO_2 \qquad (4.6)$$

The carboxylate salts of the metals Mg to Ba of the type $M(O_2CR)_2$ are all normal salts; however, reaction of $Be(OH)_2$ with carboxylic acids gives basic carboxylates, $Be_4O(O_2CR)_6$. These have a molecular structure containing a central oxygen tetrahedrally surrounded by four Be atoms, with each edge of the Be_4 tetrahedron bridged by a carboxylate ligand (Figure 4.3). Zinc (see Section 11.1) shows similar behaviour.

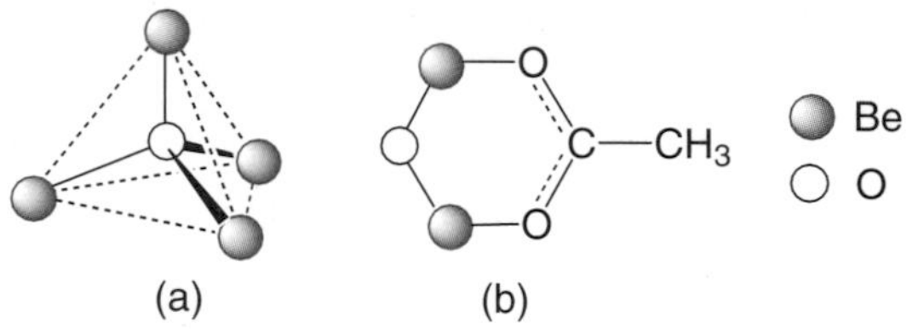

Figure 4.3 The structure of $[Be_4O(O_2CCH_3)_6]$: (a) tetrahedral arrangement of four Be atoms around the central O atom; (b) bridging of one of the Be–O–Be units by the CH_3CO_2 group

4.4   Compounds with Oxygen and Sulfur

Combustion of the Group 2 metals in oxygen gives the monoxides MO; SrO and BaO will absorb oxygen under pressure to give peroxides MO_2. However, the peroxides (and superoxides) are not formed by combustion and tend to be unstable, because the smaller M^{2+} ions are highly polarizing and cause the peroxide and superoxide salts to decompose to the monoxides MO, which have high lattice energies (see Section 3.4). This is nicely illustrated by the melting points of the MO compounds, which increase with increased lattice energy (Table 4.1). Peroxides of Mg, Ca, Sr or Ba can alternatively be made by reaction of the metal hydroxide with hydrogen peroxide (equation 4.7).

Barium peroxide is a source of hydrogen peroxide (see Section 8.3.4 and Box 8.1):
$BaO_2 + H_2SO_4 \rightarrow BaSO_4 + H_2O_2$

$$Ca(OH)_2 + H_2O_2 + 6H_2O \rightarrow CaO_2.8H_2O \qquad (4.7)$$

Table 4.1 Melting points and lattice energies of Group 2 monoxides

Oxide	Lattice energy (kJ mol^{-1})	Melting point (°C)
BeO	−4298	2507
MgO	−3800	2800
CaO	−3419	1728
SrO	−3222	1635
BaO	−3034	1475

The monoxides are *basic* (see Section 3.5) and react with water to give the hydroxides $M(OH)_2$ (equation 4.8), which can also be formed by addition of OH^- ions to solutions of M^{2+} ions (equation 4.9).

$$CaO + H_2O \rightarrow Ca(OH)_2 \qquad (4.8)$$

$$Mg^{2+} + 2OH^- \rightarrow Mg(OH)_2 \qquad (4.9)$$

$Ba(OH)_2$ is the most stable and most soluble hydroxide.

Saturated solutions of $Ca(OH)_2$ ['lime water', containing 1.5×10^{-4} mol L^{-3} $Ca(OH)_2$] are commonly used to test for carbon dioxide gas, through the formation of a white precipitate of calcium carbonate:
$Ca(OH)_2 + CO_2 \rightarrow CaCO_3 + H_2O$

Worked Problem 4.3

Q How would you prepare (a) $Mg(MeCO_2)_2$ and (b) $SrSO_4$?

A (a) React a basic magnesium salt [*e.g.* $MgCO_3$ or $Mg(OH)_2$] with ethanoic acid:

$$Mg(OH)_2 + 2MeCO_2H \rightarrow Mg(MeCO_2)_2 + H_2O$$

(b) Add a solution of any soluble sulfate salt (*e.g.* Na_2SO_4) to any soluble Sr^{2+} salt [*e.g.* $Sr(NO_3)_2$]; $SrSO_4$ is only sparingly soluble and will precipitate:

$$Sr^{2+}_{(aq)} + SO_4^{2-}_{(aq)} \rightarrow SrSO_{4(s)}$$

4.5 Compounds with Nitrogen

All of the Group 2 metals form the nitride (M_3N_2) on heating the metal and nitrogen; the nitrides react with water to produce ammonia, *e.g.* equations 4.10 and 4.11.

$$3Mg + N_2 \rightarrow Mg_3N_2 \qquad (4.10)$$

$$Mg_3N_2 + 6H_2O \rightarrow 3Mg(OH)_2 + 2NH_3 \qquad (4.11)$$

Recall from Chapter 3 that lithium is the only alkali metal to give a nitride, Li_3N.

The Group 2 metals dissolve in liquid ammonia to form blue solutions containing solvated electrons (compare the analogous behaviour of the alkali metals, Section 3.6); however, their solubilities are much lower.

4.6   Hydrides

In this reaction, the hydrogen on beryllium is coming from one of the CH_3 groups of the *tert*-butyl group, which forms 2-methylpropene.

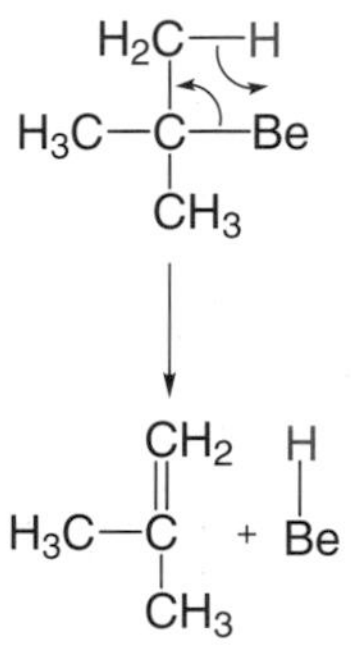

All of the Group 2 metals except beryllium form an ionic dihydride, MH_2, on heating the metal in hydrogen. BeH_2, which is stable, must be synthesized indirectly by pyrolysis of di-*tert*-butylberyllium (Bu^t_2Be), which itself is made from $BeCl_2$ and the Grignard reagent (see Section 4.7) Bu^tMgCl (equation 4.12).

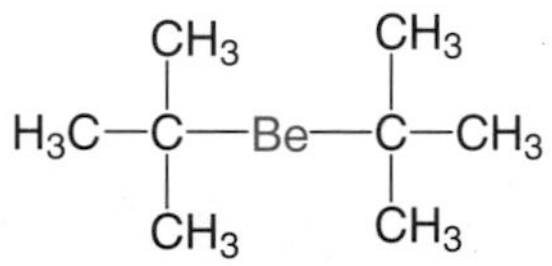

Di-*tert*-butylberyllium

$$BeCl_2 + 2Bu^tMgCl \rightarrow Bu^t_2Be \ (+\ 2MgCl_2) \rightarrow BeH_2 + 2Me_2C{=}CH_2$$
(4.12)

BeH_2 is a polymeric solid (Figure 4.4) containing Be–H–Be bridges which are two-electron, three-centre bonds, very similar in type to those in diborane (B_2H_6) and other boranes (see Section 5.6).

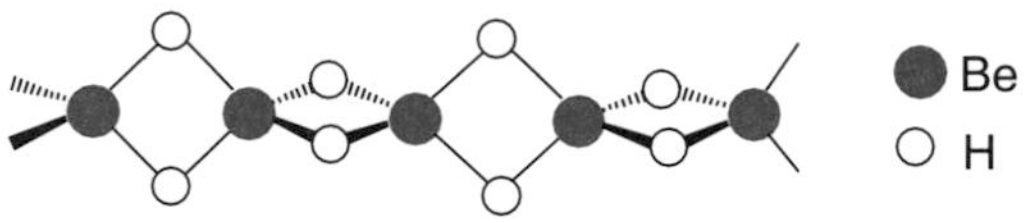

Figure 4.4 Solid-state structure of BeH_2 showing three-centre, two-electron Be⋯H⋯Be bridges

● Be

○ H

4.7   Compounds with Carbon

Carbides are classified by the type of hydrocarbon they produce on hydrolysis; see Section 6.6.

The Group 2 metals form several different types of carbides. Mg, Ca, Sr and Ba form ethynides (acetylides) of the type MC_2; an example is calcium ethynide, CaC_2, the structure of which is given in Figure 6.6. These ethynides contain the C_2^{2-} ion. Beryllium and magnesium also form two other types of carbides. Beryllium forms Be_2C on direct combination of Be and C, and the main hydrolysis product is CH_4, suggesting it contains isolated C atoms.

Worked Problem 4.4

Q Reaction of a magnesium carbide with water gave propyne (MeC≡CH). Suggest a formulation of the carbide, and give an example of a common gaseous molecule with which the carbide ion is isoelectronic.

A The carbide will contain no hydrogen atoms, so removing all the hydrogens from propyne we get the C_3^{4-} ion. Thus, the carbide will have the composition Mg_2C_3. In isoelectronic terms, a C^{2-} unit is isoelectronic with an oxygen atom (since O is in Group 16 and C in Group 14). Thus, the C_3^{4-} ion is isoelectronic with carbon dioxide, CO_2.

The most important organometallic compounds of the Group 2 metals are the **Grignard reagents**, RMgX, where X is a halide.[1] They are very similar to organolithiums, and are formed from the halide and magnesium metal in a solvent such as diethyl ether (equation 4.13).

$$R–Cl + Mg \rightarrow RMgCl \qquad (4.13)$$

Grignard reagents are solvated by the ether, giving four-coordinate Mg centres (Figure 4.5).[1] Grignard reagents are widely used as reagents for the formation of carbon–carbon bonds, *via* addition to carbonyl groups, for example equation 4.14, and for the transfer of R groups to other heteroatoms by halide displacement, *e.g.* equation 4.15.

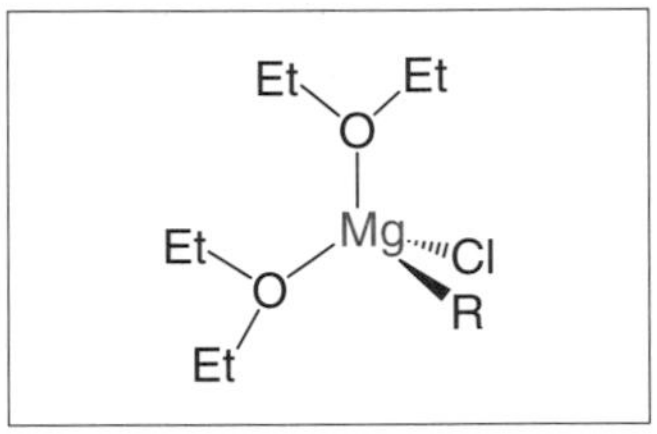

Figure 4.5 A Grignard reagent, RMgCl, solvated by diethyl ether

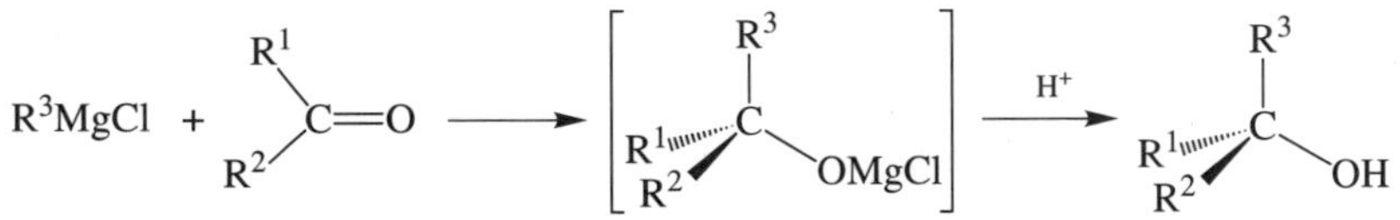

$$(4.14)$$

$$PCl_3 + 3EtMgBr \rightarrow PEt_3 + 3MgBrCl \qquad (4.15)$$

The **organomagnesium** compounds R_2Mg are known, but much less studied than the Grignard reagents. Dimethylberyllium (and also dimethylmagnesium) have the polymeric structures shown in Figure 4.6, analogous to those of solid beryllium hydride (Figure 4.4) and beryllium chloride (Figure 4.1). The sp^3 hybridized methyl group bridges by forming three-centre, two-electron electron-deficient bonds to the sp^3 hybridized metal atoms (Figure 4.7).

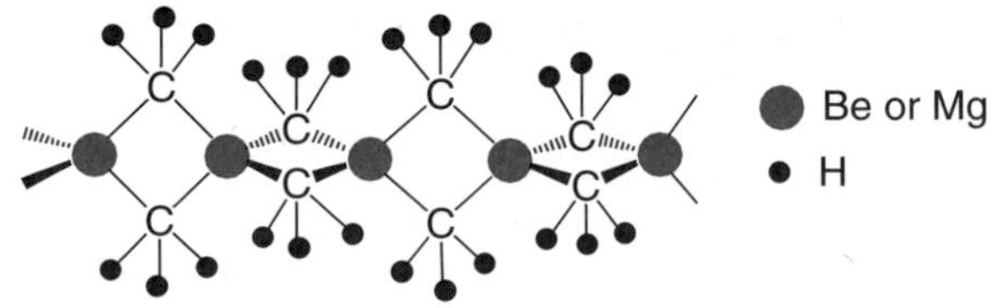

● Be or Mg
• H

Figure 4.6 Solid-state structure of $Mg(CH_3)_2$ [and $Be(CH_3)_2$]

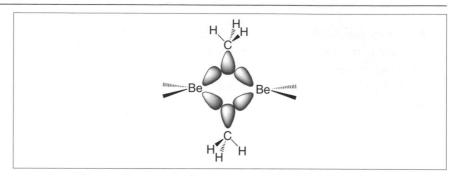

Figure 4.7 Three-centre, two-electron bonds in solid Be(CH₃)₂

Worked Problem 4.5

Q Using a Grignard reagent, how would you convert: (a) bromo-benzene into benzoic acid; (b) bromobenzene into triphenylphos-phine, PPh_3 (see Box 7.2); (c) bromoethane into ethane?

A (a) Here, a C–C bond is generated, which is done by the reaction of a Grignard reagent with a carbonyl (C=O) compound (in this case CO_2):

$$PhBr + Mg \rightarrow PhMgBr \xrightarrow[\text{2. H}^+]{\text{1. CO}_2} PhCO_2H$$

(b) Convert the PhBr into PhMgBr as in (a), then react with PCl_3:

$$PCl_3 + 3PhMgBr \rightarrow PPh_3 + 3MgBrCl$$

(c) Convert the EtBr into the Grignard reagent EtMgBr by reaction with Mg, then react with water:

$$EtMgBr + H_2O \rightarrow EtH \text{ (ethane)} + MgBr(OH)$$

Water is a stronger acid than ethane.

4.8   Group 2 Metal Complexes

Owing to their having a higher charge density than the alkali metals, the Group 2 metal ions form many more coordination complexes, though the tendency to form complexes decreases going down the group because of the larger ionic size and reduced strength of ion–dipole interactions. All of the M^{2+} cations form complexes with oxygen and nitrogen donor ligands, and with the lighter halides preferentially. Beryllium, owing to its small size, is restricted to a maximum coordination number of four, but magnesium and calcium are commonly six coordinate, and strontium and barium can have even higher coordination numbers.

4.9 Similarities in the Chemistry of Beryllium and Aluminium

The diagonal relationship (see Section 3.11) is also illustrated by these two elements, which share much common chemistry. The Al^{3+} ion is larger than Be^{2+}, so both have similar, high charge densities, and the free cations do not exist. Instead, both beryllium and aluminium form covalent compounds or strongly solvated cations which are acidic and readily hydrolysed. Both metals dissolve in non-oxidizing acids (such as HCl) or alkalis with liberation of hydrogen gas, and both form polymeric hydrides, chlorides and alkyls.

Worked Problem 4.6

Q Both beryllium and aluminium form hydrated cations. What is the major difference between the hydrated cations?

A The major difference between beryllium and aluminium is due to absolute size: aluminium, being larger, forms six-coordinate $[Al(H_2O)_6]^{3+}$ while beryllium forms four-coordinate $[Be(H_2O)_4]^{2+}$.

Summary of Key Points

1. The *formation of 2+ cations* dominates the chemistry of the Group 2 metals; all have low first and second ionization energies and electronegativities.

2. The *small size of Be^{2+}* (and to a much lesser extent Mg^{2+}) means that its compounds are either largely covalent or contain solvated ions. The Group 2 metals form more, and more stable, coordination complexes compared to the Group 1 metals.

3. All Group 2 metals form *basic oxides*, but the unique character of beryllium (owing to its small size) means that BeO is amphoteric.

Problems

4.1. What is the oxidation state of the Group 2 element in the following substances: (a) Mg metal; (b) CaC_2; (c) $[Be(H_2O)_4]^{2+}$.

4.2. Identify the Group 2 element X in each of the following:
(a) The sulfate XSO_4 is highly insoluble in water, and X forms the most soluble hydroxide $X(OH)_2$.
(b) The chemistry of X is quite different from that of the other elements in the group.
(c) X forms organometallic compounds RXBr, which are very widely used.

4.3. Explain why both $BeCO_3$ and $BeSO_4$ are unstable to heat, while $BaSO_4$ and $BaCO_3$ are stable.

4.4. Explain why all of the Group 2 metals form a nitride M_3N_2, but of the Group 1 metals only lithium forms a nitride, Li_3N.

4.5. Predict the outcome of the following reactions, and write balanced equations:
(a) $BaO_{2(s)} + H_2SO_{4(aq)} \rightarrow$
(b) $Ba(NO_3)_{2(aq)} + Na_2SO_{4(aq)} \rightarrow$
(c) $Ca_{(s)} + H_{2(g)} + heat \rightarrow$

4.6. Using a suitable Grignard reagent, how would you prepare:
(a) MeC(Et)(OH)Ph; (b) $AsPh_3$.

Reference

1. F. Bickelhaupt, *J. Organomet. Chem.*, 1994, **475**, 1.

Further Reading

Strontium – a neglected element, J. W. Nicholson and L. R. Pierce, *Educ. Chem.*, 1995, May, 74.
To be or not to Be – the story of beryllium toxicity, D. N. Skilleter, *Chem. Br.*, 1990, 26.
Beryllium coordination chemistry, C. Y. Yong and J. D. Woollins, *Coord. Chem. Rev.*, 1994, **130**, 243.

5

The Group 13 Elements: Boron, Aluminium, Gallium, Indium and Thallium

Aims

By the end of this chapter you should understand:

- The wide variation of chemical and physical properties in this group
- The tendency of the lighter elements to form electron-deficient compounds, and the tendency of boron to form polyhedral clusters
- The use of Wade's rules in rationalizing the structures of main group clusters
- The inert pair effect

12	13	14
	B	C
	Al	Si
Zn	Ga	Ge
Cd	In	Sn
Hg	Tl	Pb

5.1 Introduction and Oxidation State Survey

In Group 13, boron is typically regarded as a non-metallic element, though with some metallic characteristics. The other elements are all metals. This is illustrated by the considerably higher first, second and third ionization energies (IEs) of boron compared with the other elements (Figure 5.1). There are many similarities but also many differences in the chemistry of boron and the other elements. Aluminium has many similarities to beryllium (see Section 4.9).

For the Group 13 elements, +3 is the stable oxidation state; however, we see the stabilization of a lower oxidation state (+1) for the heaviest element in the group, thallium, for which the +3 state is oxidizing. This tendency (the inert pair effect) becomes even more pronounced as we move to the elements in Groups 14, 15 and 16.

In the **inert pair effect**, a pair of electrons occupies a low-energy s-orbital and the electrons are harder to ionize, and the lower strength of bonds involving these heavy atoms cannot compensate for the energy needed to promote the electrons. The effect is important for heavy p-block elements in Groups 13–16 (Tl, Sn, Pb, Sb, Bi, Te). See also Section 6.1.

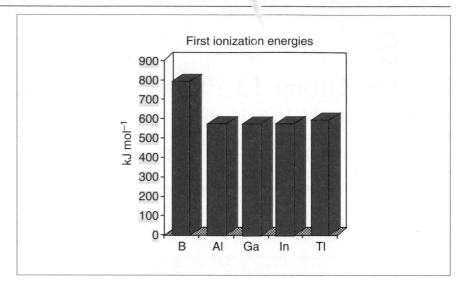

Figure 5.1 First ionization energies (kJ mol⁻¹) for the Group 13 elements

5.2   The Elements

Boron occurs naturally as **borax**, $Na_2B_4O_7.4H_2O$ and $Na_2B_4O_7.10H_2O$. Acidification of this gives **boric acid**, $B(OH)_3$ (also written as H_3BO_3), which is dehydrated to boric oxide, B_2O_3 (Section 5.7.1), by heating. Amorphous (non-crystalline) boron is then obtained by reduction of the B_2O_3 using sodium or magnesium. Crystalline boron is obtained by the reduction of a mixture of boron trihalide (*e.g.* BCl_3) and hydrogen over a heated tantalum wire (equation 5.1). Boron crystallizes in a variety of forms, all containing the icosahedral B_{12} unit (Figure 5.2). The propensity of boron to form such polyhedral clusters is a commonly occurring theme of the chemistry of this element in low oxidation state compounds, especially hydrides (Section 5.6) and some halides (Section 5.5.2).

$$2BCl_{3(g)} + 3H_{2(g)} \rightarrow 2B_{(s)} + 6HCl_{(g)} \qquad (5.1)$$

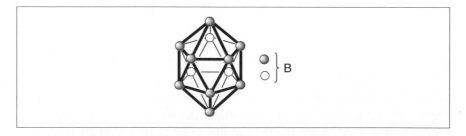

Figure 5.2 The icosahedral B_{12} unit found in crystalline boron (and some metallic borides)

After oxygen and silicon, **aluminium** is the most common element in the Earth's crust (8.1%), where it can replace silicon in silicates, giving aluminosilicates (see Section 12.3). The most important ore of aluminium is *bauxite*, $Al_2O_3.H_2O$, which is purified by dissolving it in strong NaOH solution, removing the insoluble iron impurities, and reprecipitating the aluminium as $Al_2O_3.3H_2O$ (hydrated alumina). The metal is obtained by electrolysing the dried alumina in molten Na_3AlF_6.

The other three elements, **gallium**, **indium** and **thallium**, are of fairly low abundance, occurring as minor components of a range of minerals, such as *sphalerite* (ZnS) (see Section 11.5).

5.3   Chemistry of the Elements

Elemental boron combines with oxygen, halogens, sulfur and nitrogen, and with many metals. It is resistant to acids, and only reacts with molten sodium hydroxide above 500 °C.

Aluminium, while being a very reactive metal, is usually made unreactive ('passivated') by a thin coating of aluminium oxide. Aluminium dissolves in HCl to give the $[Al(H_2O)_6]^{3+}$ ion plus hydrogen, and in strong hydroxide solutions, giving aluminates and hydrogen (equation 5.2).

$$2Al + 2NaOH + 6H_2O \rightarrow 2NaAl(OH)_4 + 3H_2 \qquad (5.2)$$

Gallium, indium and thallium are reasonably reactive metals which readily dissolve in acids to give trivalent cations for Ga and In but Tl^+ for thallium (equations 5.3 and 5.4), illustrating the greater stability of the +1 oxidation state for this element.

$$2In_{(s)} + 6HCl_{(aq)} \rightarrow 2InCl_{3(aq)} + 3H_{2(g)} \qquad (5.3)$$

$$2Tl_{(s)} + 2HNO_{3(aq)} \rightarrow 2TlNO_{3(aq)} + H_{2(g)} \qquad (5.4)$$

5.4   Borides

When boron is heated with most metals, metal borides are formed; in this aspect of its chemistry, boron shows similarities with carbon and silicon, which form carbides and silicides, respectively. The structures of these borides are dependent on the metal-to-boron ratio, and contain single, pairs, chains, double chains, sheets or clusters of boron atoms. Compounds with composition M_2B (*e.g.* Fe_2B) have single boron atoms, while those with a 1:1 ratio (*e.g.* FeB) have single chains of boron atoms running through the metal lattice (Figure 5.3). In materials of composition MB_2 (*e.g.* MgB_2, TiB_2) the material has a sheet structure (Figure 5.3), while in MB_6 (*e.g.* MgB_6) there are clusters of six boron atoms contained within a cube of metal atoms, in a caesium chloride-type structure (see Figure 3.3). Finally, in compounds of composition MB_{12} (*e.g.*

AlB$_{12}$) the boron atoms form a network of linked icosahedral B$_{12}$ clusters (Figure 5.2), as found in crystalline boron itself.

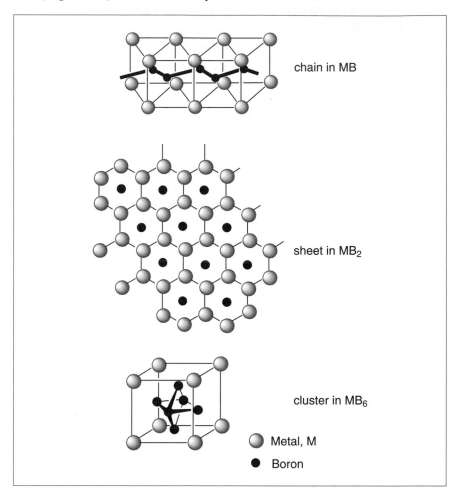

chain in MB

sheet in MB$_2$

cluster in MB$_6$

○ Metal, M

● Boron

Figure 5.3 The structures adopted by some metal borides

5.5   Halides

5.5.1   Trivalent Halides, MX$_3$

All combinations of M and X occur in the trivalent MX$_3$ halides, with the exception of thallium(III) iodide.

The boron trihalides BX$_3$ are planar molecules, having an empty p-orbital perpendicular to the molecular plane. In BF$_3$ (and to a far lesser extent in BCl$_3$) π back-donation from filled halogen p-orbitals into the empty boron p-orbital can occur (Figure 5.4). Back donation is effective for fluorine, with small p-orbitals, which gives very effective overlap, and essentially some partial multiple π-bond character in the B–F bonds. For

BCl_3, overlap with the larger chlorine orbitals is poorer (Figure 5.4), so the boron is more electron deficient and a more powerful **Lewis acid** than BF_3.

A **Lewis acid** is an electron-pair **acceptor**, whereas a **Lewis base** is an electron-pair **donor**.

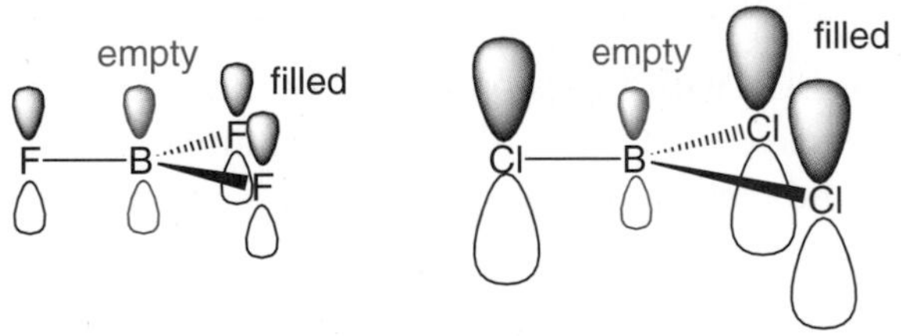

Figure 5.4 Back donation from filled halogen p-orbitals into an empty boron p-orbital in the halides BX_3

Worked Problem 5.1

Q Classify the following species as Lewis acids or Lewis bases: (a) H^+; (b) diethyl ether; (c) Ph_3Si^+.

A (a) The H^+ is a naked proton, with no electrons, so it can only act as a Lewis acid. (b) Diethyl ether, Et_2O, has an oxygen with two lone pairs, so it is a Lewis base. (c) The Ph_3Si^+ cation has a three-coordinate silicon atom which is electron deficient, so it will act as a Lewis acid.

While all of the boron trihalides BX_3 are monomers, the structures of the AlX_3 compounds are dependent on the halide X. AlF_3 is a high-melting polymeric solid built from fluoride-bridged AlF_6 octahedra. The structure of $AlCl_3$ in the solid state similarly has six-coordinate Al centres, with chloride bridges. However, in the liquid and gas phases the dimer Al_2Cl_6 (**5.1**) is formed, where one of the chlorides forms a dative bond to the other aluminium centre, completing its octet. $AlBr_3$ and AlI_3 are dimeric in all states.

All the trihalides MX_3 are powerful Lewis acids, forming adducts of the type MX_3L. Indeed, BF_3 is often used in this form, as an adduct (**5.2**) with diethyl ether. The formation of anions MX_4^- from MX_3 by addition of halide can also be viewed as Lewis acid–Lewis base complex formation, e.g. the formation of BF_4^- from BF_3. For aluminium and the other heavier group members, more than one ligand is able to be added, up to a maximum of six coordination.

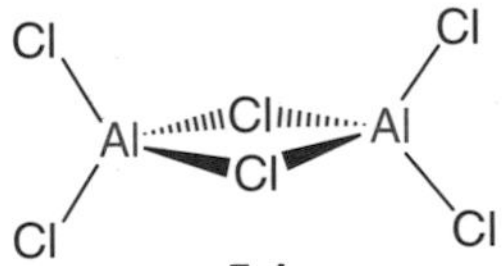

5.1

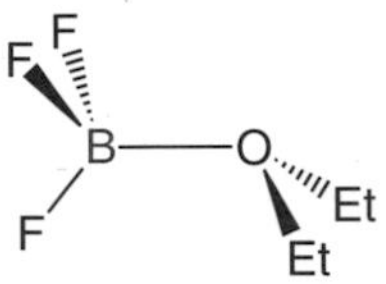

5.2 Structure of $BF_3.Et_2O$

Examples of halide complexes:
AlF_6^{3-} (octahedral)
$InCl_5^{2-}$ (both trigonal-bipyramidal and square-pyramidal forms exist)
$GaBr_4^-$ (tetrahedral)

Worked Problem 5.2

Q Boron trichloride is a volatile liquid (boiling point 12.5 °C). Suggest a way in which it can be more conveniently handled.

A Adducts of BCl_3 with Lewis bases will be much less volatile than BCl_3 itself. Thus, the adduct with dimethyl sulfide, $Me_2S.BCl_3$, is a crystalline solid, which is easy to handle, and is less sensitive towards hydrolysis than BCl_3.

Worked Problem 5.3

Q Predict the outcome of the following reactions:
(a) $BF_3.NMe_3 + BCl_3 \rightarrow$
(b) $BCl_3.NMe_3 + BF_3 \rightarrow$

A BCl_3 is a stronger Lewis acid than BF_3; thus in reaction (a) it will displace it to form $BCl_3.NMe_3$ and free BF_3. In (b) there will be no reaction.

5.5.2 Lower Oxidation State Halides

All the Group 13 elements form diatomic halides MX, although all except thallium are unstable towards disproportionation to the metal and the trivalent halide (even gaseous TlCl is unstable to disproportionation). As an example, AlCl and GaCl can be formed by the reaction of aluminium or gallium metal with HCl gas at high temperature and low pressure, giving red AlCl or GaCl, which are condensed at low temperature (77 K). On warming, they **disproportionate** (equation 5.5).

Disproportionation is where some atoms of the same element are oxidized and others reduced in the same reaction.

$$3MCl \rightarrow MCl_3 + 2M \tag{5.5}$$

5.3

Boron is notable for lower oxidation state halides containing B–B bonds. Thus B_2Cl_4 (**5.3**) is made by passing an electrical discharge through BCl_3 using mercury electrodes (equation 5.6), or by condensing copper atoms with BCl_3 (equation 5.7). B_2F_4, B_2Br_4 and B_2I_4 are also known.

$$2BCl_3 + 2Hg \rightarrow B_2Cl_4 + Hg_2Cl_2 \tag{5.6}$$

$$2BCl_3 + 2Cu(atoms) \rightarrow B_2Cl_4 + 2CuCl \tag{5.7}$$

Boron forms a number of halides which contain clusters of boron atoms, with each boron bearing a halogen atom. B_4Cl_4 is a by-product of reaction 5.6, while B_8Cl_8 and B_9Cl_9 (and the larger clusters $B_{10}Cl_{10}$, $B_{11}Cl_{11}$ and $B_{12}Cl_{12}$) are formed progressively by decomposition of B_2Cl_4 at room temperature. The structures of B_4Cl_4 and B_9Cl_9 are shown in Figure 5.5.

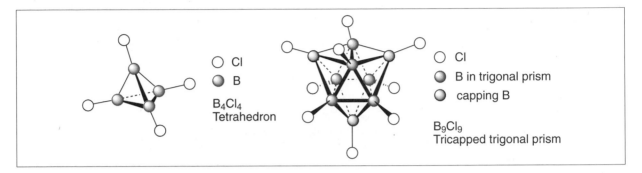

○ Cl
● B

B_4Cl_4
Tetrahedron

○ Cl
● B in trigonal prism
○ capping B

B_9Cl_9
Tricapped trigonal prism

Gallium(II) occurs in the species $Ga_2X_6^{2-}$ (X = Cl, Br or I), formed by electrolysis of gallium metal in strong acid. These contain a Ga–Ga bond (thus accounting for the +2 oxidation state), but are readily oxidized, *e.g.* by halogens X_2 to GaX_4^-.

Figure 5.5 Structures of some boron chloride clusters

5.6   Hydrides and Organometallic Compounds

5.6.1 Boron Hydrides

Boron forms more hydrides than any other Group 13 element; there are very extensive series of *electron-deficient* hydride compounds, where the usual two-centre, two-electron covalent bonds (formed in hydrides such as PH_3 and H_2S) are supplemented by three-centre, two-electron bonds. Some of the better-known boranes are listed in Table 5.1; the compounds fall into two types, having compositions B_nH_{n+4} and B_nH_{n+6}.

The simplest possible borane is BH_3, but this compound has never been isolated. Instead, the smallest borane is the dimer B_2H_6, **diborane**, formed by the reduction of a boron trihalide with $Li[AlH_4]$ (equation 5.8).

$$4BF_3 + 3Li[AlH_4] \rightarrow 2B_2H_6 + 3Li[AlF_4] \qquad (5.8)$$

B_2H_6 has the structure shown in Figure 5.6, with four terminal B–H bonds (which are normal two-electron covalent bonds), and two three-centre, two electron B⋯H⋯B bonds. Each boron atom is sp^3 hybridized; two hybrids are used to bond the terminal H atoms, while the other two form three-centre bonds with hydrogen 1s orbitals, as shown in Figure 5.7.

Table 5.1 Some boron hydrides

B_nH_{n+4}	B_nH_{n+6}
B_2H_6	B_4H_{10}
B_5H_9	B_5H_{11}
B_6H_{10}	B_6H_{12}
B_8H_{12}	B_8H_{14}
$B_{10}H_{14}$	

The electron-deficient bonding in B_2H_6 is similar to that in solid $BeMe_2$ (see Chapter 4), $AlMe_3$ and $(AlH_3)_x$ (Section 5.6.2).

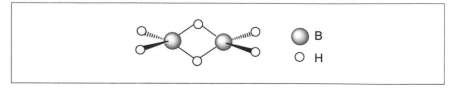

Figure 5.6 The structure of diborane, B_2H_6

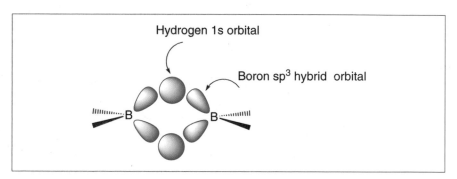

Figure 5.7 Three-centre, two-electron bonds in B_2H_6

Nomenclature: pentaborane-9 refers to the borane with five B atoms ('penta') and nine hydrogens: B_5H_9.

Higher boron hydrides contain the same structural features as B_2H_6, with one or more B–B bonds present. The molecules have 'open' polyhedral cluster structures; the structures of tetraborane (B_4H_{10}) and pentaborane-9 (B_5H_9) are shown in Figure 5.8. These higher boron hydrides can be formed by heating B_2H_6; different reaction conditions give different boron hydrides, for example equations 5.9 and 5.10.

$$2B_2H_6 \xrightarrow{100-120°C} B_4H_{10} + H_2 \qquad (5.9)$$

$$5B_2H_6 \xrightarrow{180-220°C} 2B_5H_9 + 6H_2 \qquad (5.10)$$

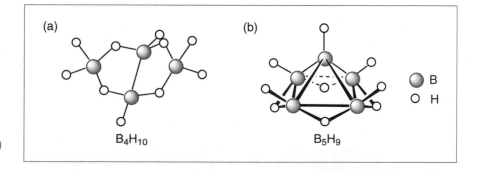

(a)   (b)

B_4H_{10} B_5H_9

Figure 5.8 The structures of (a) B_4H_{10} and (b) B_5H_9

Worked Problem 5.4

Q A borane contains 84.2% boron and 15.7% hydrogen, by elemental analysis, and has a relative molecular mass of 76.7. Determine the molecular formula.

A Divide percentages by appropriate atomic masses, to get molar ratios:

B: 84.2/10.811 = 7.788; H: 15.7/1.0079 = 15.58; therefore H/B ratio is 2.001. Hence the compound is B_6H_{12}, which has the correct molecular mass (and is a well-known borane).

Worked Problem 5.5

Q Hydrides of the Group 13 elements are potentially suitable for use as rocket fuels because they produce a large amount of energy on combustion. How much energy is available from the complete combustion of 1 mole of B_6H_{10}? Enthalpies of formation: $\Delta_f H°$: B_6H_{10}, +56 kJ mol^{-1}; B_2O_3, –1273 kJ mol^{-1}; H_2O, –286 kJ mol^{-1}.

A ΔH for a reaction is the sum of the $\Delta_f H°$ values for the products, minus the sum of the $\Delta_f H°$ values of the reactants, remembering $\Delta_f H°$ of an element in its standard state is zero. Firstly, write a balanced equation for the combustion:

$$B_6H_{10} + 7O_2 \rightarrow 3B_2O_3 + 5H_2O$$

ΔH for the reaction is then $3\Delta_f H°(B_2O_3) + 5\Delta_f H°(H_2O) - \Delta_f H°(B_6H_{10}) - \Delta_f H°(O_2) = 3(-1273) + 5(-286) - (56) - (0) = -5305$ kJ mol^{-1}.

As well as the neutral boranes, there is an extensive series of **boron hydride anions**, which have closed polyhedral skeletons. Thus, $[B_6H_6]^{2-}$ and $[B_{12}H_{12}]^{2-}$ have octahedral B_6 and icosahedral B_{12} cages, respectively, and terminal B–H bonds, as shown in Figure 5.9. **Carboranes** are obtained by replacing one or more BH units with C atoms.

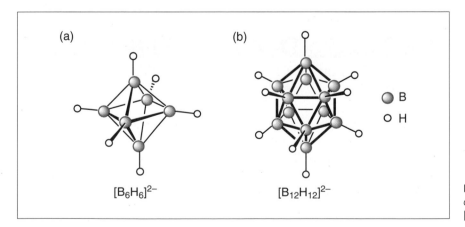

(a)   (b)

B
H

$[B_6H_6]^{2-}$ $[B_{12}H_{12}]^{2-}$

Figure 5.9 The structures of the *closo* boron hydride anions (a) $[B_6H_6]^{2-}$ and (b) $[B_{12}H_{12}]^{2-}$

Wade's rules can be applied to a wide range of other cluster species of main group and transition metals.

The shapes of the various borane clusters can be rationalized using **Wade's rules**, which consider the total number of **skeletal electron pairs (SEP)** available for cluster bonding.

Box 5.1 Wade's Rules

A borane cluster can be built up from BH units (with terminal, *i.e.* non-bridging, hydrogens). Other hydrogens in the cluster bond in other ways. The number of SEP is calculated as follows.

1. Determine the total number of valence electrons (three per B atom, one per H atom), adding or subtracting electrons for any overall charge (a C atom in a carborane contributes four electrons).

2. Subtract two electrons for each BH (or CH) unit.

3. The number of electrons, divided by 2, is the number of SEP, which determines the type of cluster.

4. A cluster with n vertices occupied by n B atoms requires $n+1$ SEP for bonding. Such clusters are termed *closo*; examples are the octahedron and icosahedron (Figure 5.11), with 7 and 13 SEP, respectively.

5. If there are $n+1$ SEP and $n-1$ B atoms, the cluster is derived from a *closo* cluster, but with one missing vertex: this is called a *nido* cluster.

6. If there are $n+1$ SEP and $n-2$ B atoms, the cluster has two missing vertices and is called an *arachno* cluster. The simple theory does not say unequivocally which vertices are missing.

7. If there are more than n B atoms and $n+1$ SEP, the extra B atoms occupy *capping positions* over triangular faces.

Worked Problem 5.6

Q Confirm that the predicted shape of pentaborane-9 is as shown in Figure 5.8.

A Using the principles in Box 5.1, first, count the number of valence electrons: 5B atoms provide 15 electrons, and 9 hydrogens provide 9 electrons, giving 24. There are 5 BH units, so subtracting 2 electrons for each (10 total) gives 14 electrons or 7 SEP. An octahedron (with 6 vertices) requires (6+1) or 7 SEP; hence the shape of B_5H_9 is derived from an octahedron. However, there are only 5 B atoms, so the structure is a *nido* cluster: an octahedron with 1 missing vertex (*i.e.* a square pyramid). The shape of B_5H_9 in Figure 5.8 is also a square pyramid.

Worked Problem 5.7

Q Using Wade's rules, predict the structure of B_5H_{11}.

A Total electron count = $(5 \times 3) + (11 \times 1) = 26$. Subtract (5×2) electrons for 5 BH units, giving 16 electrons, or 8 pairs, so the structure is based on a seven-vertex solid (a pentagonal bipyramid). Since there are only 5 B atoms, 2 vertices are missing, and the cluster is of the *arachno* type. Wade's rules do not say which vertices are missing; the missing vertices and the structure of B_5H_{11} are shown in Figure 5.10.

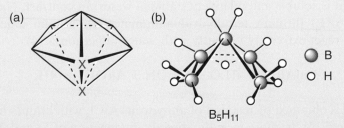

Figure 5.10 (a) A pentagonal bipyramid, showing the positions of the two missing vertices (X) in B_5H_{11}; (b) the structure of B_5H_{11}

5.6.2 Hydrides of the Other Group 13 Elements

In contrast to boron, few other hydrides are known. For aluminium, the only stable hydride is polymeric **alane** $(AlH_3)_x$. For gallium, the reaction of the dimeric $[H_2GaCl]_2$ with $Li[GaH_4]$ gives **gallane** $[GaH_3]_n$, which mainly occurs as the rather thermally unstable diborane-like dimer Ga_2H_6 in the gas phase.[1] In some respects the chemistry of gallane more resembles that of borane than alane, because the electronegativity of gallium (1.8) is higher than that of aluminium (1.5) and closer to boron (2.0).

5.6.3 Hydride Adducts

A common example of a donor (D) is an amine (*e.g.* Me_3N), as in $Me_3N.GaH_3$.

The hydrides of the Group 13 elements form donor-stabilized monomers, $MH_3.D$ (where D is the donor), as illustrated by the formation of BH_3 adducts (equation 5.11).

$$B_2H_6 + 2D \rightarrow 2BH_3.D \qquad (5.11)$$

Salts containing the tetrahedral MH_4^- ions are obtained by various reactions, including where the donor D in equation 5.14 is a hydride ion, H^-. Sodium tetrahydridoborate(III), $Na[BH_4]$ (commonly called sodium borohydride), is produced from trimethyl borate (see Section 5.7.1) and sodium hydride (equation 5.12), and is widely used as a reducing agent for carbonyl groups, *e.g.* equation 5.13.

$$B(OMe)_3 + 4NaH \rightarrow Na[BH_4] + 3NaOMe \qquad (5.12)$$

$$Me_2CO + Na[BH_4] \rightarrow Me_2CHOH \qquad (5.13)$$

Sodium borohydride is particularly useful as a mild reducing agent because it is soluble and relatively stable in water. In contrast, $Na[AlH_4]$ and $Li[AlH_4]$ [lithium tetrahydridoaluminate(III), lithium aluminium hydride] react extremely violently with water (equation 5.14).

$$Na[AlH_4] + 4H_2O \rightarrow NaOH + Al(OH)_3 + 4H_2 \qquad (5.14)$$

$Li[AlH_4]$ is obtained by reaction of anhydrous $AlCl_3$ with lithium hydride (equation 5.15) or, more recently, directly, by reaction of Li and Al metals with hydrogen at elevated temperature (equation 5.16).

$$4LiH + AlCl_3 \rightarrow Li[AlH_4] + 3LiCl \qquad (5.15)$$

$$Li + Al + 2H_2 \rightarrow Li[AlH_4] \qquad (5.16)$$

$Li[AlH_4]$ is a powerful reducing agent, widely used in the laboratory, *e.g.* for conversion of chlorophosphines into primary or secondary phosphines (by conversion of P–Cl bonds into P–H bonds) (equation 5.17).

$$2PhPCl_2 + Li[AlH_4] \rightarrow 2PhPH_2 + Li[AlCl_4] \qquad (5.17)$$

Worked Problem 5.8

Q Predict the outcome of each of the following reactions:
(a) $PhSiCl_3 + Li[AlH_4] \rightarrow$
(b) $Ph_2C=O + Na[BH_4] \rightarrow$
(c) $Se + Na[BH_4] \rightarrow$

A (a) $PhSiCl_3 + Li[AlH_4] \rightarrow PhSiH_3$ (phenylsilane)
(b) $Ph_2C=O + Na[BH_4] \rightarrow Ph_2CHOH$ (diphenylmethanol)
(c) $Se + Na[BH_4] \rightarrow NaHSe$ (sodium hydrogenselenide) and H_2Se

5.7   Oxides, Hydroxides and Oxyanions

All of the oxides M_2O_3 can be made by heating the element in oxygen, and are summarized in Table 5.2. Reaction of aqueous solutions of the metal trihalides with hydroxide gives the oxides in hydrated form. Going down the group, there is a transition from acidic oxides, through amphoteric, to basic, owing to increasing metallic character of the element concerned.

The same trend in acidity/basicity is observed for oxides of Groups 14–16.

Table 5.2 Oxides of the Group 13 elements

Oxide	Properties
B_2O_3	Weakly acidic
Al_2O_3	Amphoteric
Ga_2O_3	Amphoteric
In_2O_3	Weakly basic
Tl_2O_3	Basic, oxidizing

5.7.1 Boron

Boric oxide (B_2O_3) is an **acidic oxide**, readily hydrated to form boric acid, $B(OH)_3$, which is a weak acid, with a hydrogen-bonded layer structure (Figure 5.11). Its acidic nature is due to the reaction shown in equation 5.18.

Acidic oxides dissolve in water to release H+; other examples of acidic oxides are CO_2, SO_2 and P_4O_{10}.

$$B(OH)_3 + 2H_2O \rightarrow B(OH)_4^- + H_3O^+ \qquad (5.18)$$

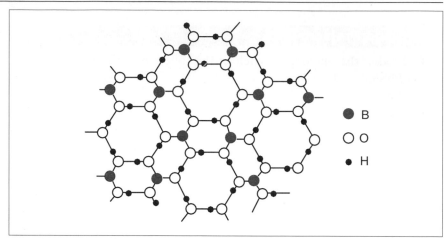

Figure 5.11 The hydrogen-bonded layer structure of boric acid, $B(OH)_3$

Boric acid reacts with alcohols to form borates, for example trimethyl borate (equation 5.19).

$$B(OH)_3 + 3MeOH \rightarrow B(OMe)_3 + 3H_2O \qquad (5.19)$$

This reaction occurs particularly readily with 1,2-dialcohols (such as ethane-1,2-diol, $HOCH_2CH_2OH$), because stable five-membered rings are formed (see structure **5.4**). There is a very extensive chemistry of **borate anions**, which are based on planar BO_3 groups or tetrahedral BO_4 groups (or both), and a vast array of compounds are known. The structures of some selected, commonly occurring borate anions are shown in Figure 5.12. Complex equilibria serve to interchange the various anions.

5.4

The simple borate anion BO_3^{3-} is isoelectronic with both the carbonate CO_3^{2-} and nitrate NO_3^- anions.

5.7.2 Aluminium

There are two forms of Al_2O_3; the high-temperature α form (which occurs as the hard mineral *corundum*) has a cubic close-packed array of oxide ions with the Al^{3+} ions regularly arranged in octahedral holes, while the lower-temperature γ form has a more complex but open structure.

Aluminium forms two hydrated oxides, $MO(OH)$ and $M(OH)_3$. Aluminum oxide is amphoteric (see Section 4.3) and dissolves in concentrated hydroxide solution to give aluminate solutions containing the $[Al(OH)_4]^-$ ion.

Gem forms of Al_2O_3:
White sapphire
Ruby (contains traces of Cr^{3+})
Blue sapphire (contains traces of Fe^{2+}, Fe^{3+} or Ti^{4+})

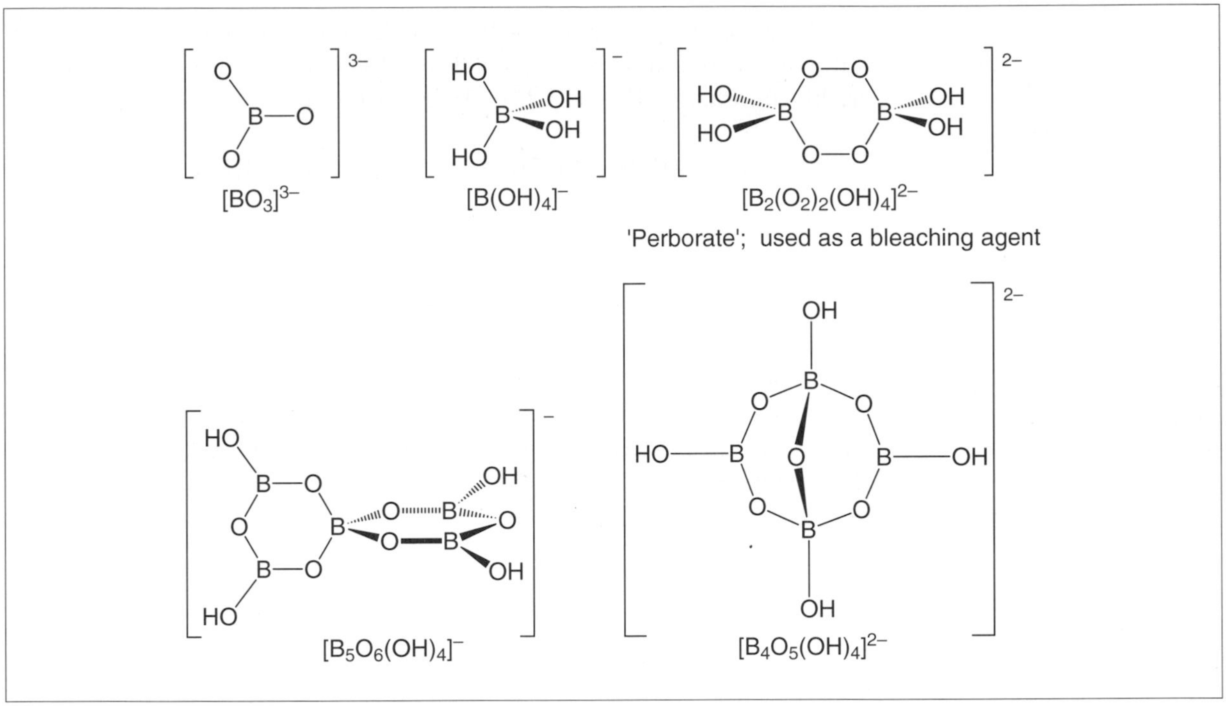

$[BO_3]^{3-}$

$[B(OH)_4]^-$

$[B_2(O_2)_2(OH)_4]^{2-}$
'Perborate'; used as a bleaching agent

$[B_5O_6(OH)_4]^-$

$[B_4O_5(OH)_4]^{2-}$

Figure 5.12 The structures of some borate anions

5.8   Compounds with Groups 15 and 16 Elements

5.8.1   Compounds with P, As, S, Se and Te

Heating the Group 13 elements (M) with elemental sulfur, selenium or tellurium (E) gives the chalcogenides M_2E_3, which contain either four- or six-coordinate M atoms. Materials such as InP and GaAs are important semiconductor materials (termed 'III–V semiconductors').

5.8.2   Compounds with N: Borazines and Boron Nitride

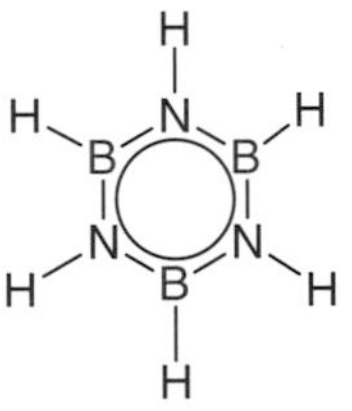

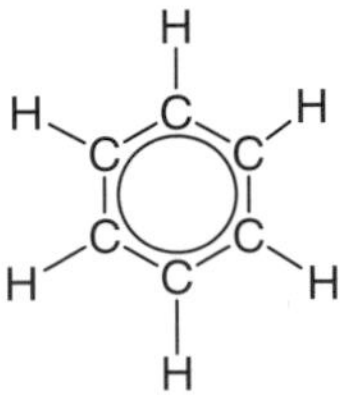

As mentioned in Section 5.6.3, BH_3 forms adducts with a wide range of donor ligands, such as $Me_3N.BH_3$. When the donor is also a hydride, loss of hydrogen can also occur. For example, B_2H_6 and NH_3 initially form the expected $H_3N.BH_3$, which gives the compound **borazine**, $B_3N_3H_6$ (**5.5**), on warming to room temperature.

Borazine is isoelectronic with benzene (C_6H_6, **5.6**) and also has a delocalized structure, with all B–N bond lengths identical (144 pm). However, in borazine the electronegative N atoms are polarized δ– and the electropositive B atoms δ+, so the reactivity of borazine is rather different to that of benzene, and the sometimes-used term 'inorganic benzene' is not a good description of borazine.

5.5 Borazine

5.6 Benzene

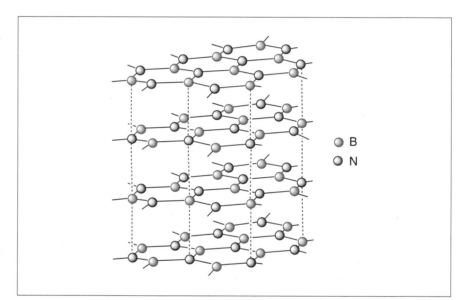

5.7 *B,B,B*-Trichloroborazine

There are many other derivatives of this type, involving different substituents on B and/or N. The reaction of NH_4Cl with BCl_3 gives *B,B,B*-trichloroborazine, $B_3N_3H_3Cl_3$ (**5.7**) (equation 5.20).

$$3NH_4Cl + 3BCl_3 \rightarrow B_3N_3H_3Cl_3 + 9HCl \qquad (5.20)$$

Just as borazine is isoelectronic with benzene, the material boron nitride (BN) is isoelectronic with carbon. Both carbon and BN form diamond- and graphite-like structures. The graphite-like structure of BN (Figure 5.13) has layers consisting of planar, hexagonal B_3N_3 rings (with BN distances similar to those in borazine). These layers stack on top of each other such that the $B^{\delta+}$ and $N^{\delta-}$ atoms lie on top of each other; this contrasts with the staggered arrangement found in graphite (see Figure 6.1) and borazine itself.

B

N

Figure 5.13 The structure of the layered form of boron nitride (BN). Alternate layers stack B···N···B···N

Worked Problem 5.9

Q Borazine reacts with three mole equivalents of HCl to give a material with chemical composition $B_3N_3H_9Cl_3$. (a) What is the structure of the product? (b) How does the isoelectronic benzene react with HCl?

A (a) The B–N bonds in borazine ($B_3N_3H_6$) are polarized $B^{\delta+}$–$N^{\delta-}$. The product is formed by addition of three moles of HCl to borazine, and the partially negatively charged nitrogen is likely to be the site of proton addition. The product formed (**5.8**) contains a saturated ring system.

(b) Benzene does not react with HCl, indicating significant differences between these two formally isoelectronic molecules.

5.8

5.9 Aquated and Related Complex Cations

The $B^{3+}_{(aq)}$ ion is unknown owing to the extreme polarizing power to be expected of a B^{3+} cation. However, salts containing the $[Al(H_2O)_6]^{3+}$ cation are known, such as potash alum, $KAl(SO_4)_2.12H_2O$. In solution the $[Al(H_2O)_6]^{3+}$ cation tends to undergo hydrolysis by loss of protons from coordinated H_2O molecules (equation 5.21).

Compare $[Be(H_2O)_4]^{2+}$ (Section 4.3).

$$[Al(H_2O)_6]^{3+} \rightleftharpoons [Al(H_2O)_5(OH)]^{2+} + H^+ \rightleftharpoons [Al(H_2O)_4(OH)_2]^+ + H^+ \rightleftharpoons$$

$$[Al(H_2O)_3(OH)_3] + H^+ \qquad (5.21)$$

The behaviour of the gallium and indium ions, $[M(H_2O)_6]^{3+}$, resembles the aluminium case, but there is a greater tendency to form complex anions if a suitable ligand, such as halide, is present, *e.g.* equation 5.22.

$$[In(H_2O)_6]^{3+} + 5Cl^- \rightarrow [InCl_5]^{2-} + 6H_2O \qquad (5.22)$$

Summary of Key Points

1. *Variation in properties:* the elements Al to Tl have low electronegativities and are metals, while boron is a non-metal, but with metalloidal tendencies. There are changes in the properties of compounds going down the group, exemplified by the oxides: B_2O_3 is an acidic oxide, Al_2O_3 and Ga_2O_3 are amphoteric, while In_2O_3 is basic.

2. The *+3 oxidation state* dominates the chemistry, but in this group the *inert pair effect* appears for the first time, in the stabilization of the +1 oxidation state (particularly for Tl).

3. *Cluster species* are formed by boron as a result of its electron deficiency; the shapes of these clusters can be rationalized by a simple electron-counting procedure (Wade's rules).

Problems

5.1. Give examples of compounds which formally contain boron, aluminium and thallium in the +1 oxidation state

5.2. Identify the element X in each of the following:
(a) X forms a chloride XCl_3 which is planar.
(b) The oxide of X is amphoteric.
(c) Addition of fluoride to XF_3 gives *only* XF_4^-.
(d) The hydride X_2H_6 has only recently been discovered, and X has a low melting point.
(e) The chloride XCl_3 is oxidizing, but XCl is stable and insoluble in water.

5.3. Explain what is meant by:
(a) The inert pair effect.
(b) An amphoteric oxide.

5.4. Predict the outcome of the following reactions, and write balanced equations:
(a) $BBr_3 + H_2O \rightarrow$
(b) $BCl_3 + Me_4N^+Cl^- \rightarrow$
(c) $Ph_2PCl + Li[AlH_4] \rightarrow$

5.5. Explain why aluminium fluoride is a solid with a high melting point, but aluminium bromide has a low melting point (97 °C) and dissolves readily in benzene.

5.6. Give an example of: (a) an electron-deficient molecule; (b) a donor–acceptor complex.

5.7. Using Wade's rules, predict the structures of the following borane clusters:
(a) $[B_6H_6]^{2-}$; (b) $B_{10}C_2H_{12}$; (c) B_4H_{10}.

5.8. Explain why an aqueous solution of aluminium(III) nitrate is acidic, but an aqueous solution of thallium(I) nitrate is not.

5.9. Amorphous boron is obtained by the reduction of B_2O_3 using metallic Mg (Section 5.2). Write a balanced equation for the reaction.

5.10. Balance equations (b) and (c) and then compare the energy available from the complete combustion of $LiBH_4$, B_4H_{10} and $B_{10}H_{14}$ (with excess dioxygen) in terms of kJ g^{-1} of the Group 13 compound.

(a) $2LiBH_4 + 4O_2 \rightarrow Li_2O + B_2O_3 + 4H_2O$

(b) $B_4H_{10} + O_2 \rightarrow$

(c) $B_{10}H_{14} + O_2 \rightarrow$

$\Delta_f H°$ values (kJ mol^{-1}): $LiBH_4$ (−189), B_2O_3 (−1273), Li_2O (−598), H_2O (−286), B_4H_{10} (+66), $B_{10}H_{14}$ (+32).

Reference

1. C. R. Pulham, A. J. Downs, M. J. Goode, D. W. H. Rankin and H. E. Robertson, *J. Am. Chem. Soc.*, 1991, **113**, 5149.

Further Reading

Boranes and metalloboranes: structure, bonding and reactivity, C. E. Housecroft, Ellis Horwood, Chichester, 1990.

Chemistry of aluminium, gallium, indium and thallium, ed. A. J. Downs, Blackie, London, 1993.

Advances in thallium aqueous solution chemistry, J. Glaser, *Adv. Inorg. Chem.*, 1995, **43**, 1.

Organoelement compounds with Al–Al, Ga–Ga and In–In bonds, W. Uhl, *Angew. Chem., Int. Ed. Engl.*, 1993, **32**, 1386.

Chemistry of the polyhedral boron halides and the diboron tetrahalides, J. A. Morrison, *Chem. Rev.*, 1991, **91**, 35.

Taking stock: the astonishing development of boron hydride cluster chemistry, N. N. Greenwood, *Chem. Soc. Rev.*, 1992, **21**, 49.

The hydrides of aluminium, gallium, indium and thallium: a re-evaluation, A. J. Downs and C. R. Pulham, *Chem. Soc. Rev.*, 1994, **23**, 175.

The hunting of the gallium hydrides, A. J. Downs and C. R. Pulham, *Adv. Inorg. Chem.*, 1994, **41**, 171.

The metallic face of boron, T. P. Fehlner, *Adv. Inorg. Chem.*, 1990, **35**, 199.

6

The Group 14 Elements: Carbon, Silicon, Germanium, Tin and Lead

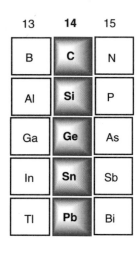

Aims

By the end of this chapter you should understand:

- The periodic changes among the chemistry of these elements, and differences with elements in neighbouring groups
- The stability and abundance of compounds containing C–C bonds
- The importance of π-bonding in carbon compounds

6.1 Introduction and Oxidation State Survey

The Group 14 elements lie in the centre of the main group elements (if the Group 1 and Group 2 elements are included), and they show major differences from the lightest element carbon (a typical non-metal) to the heaviest element lead (a typical main group metal). The group oxidation state is +4, and for C and Si this is an important and stable state. In hydrocarbons the oxidation state of carbon is formally –4; these compounds are *thermodynamically* unstable, but an extremely important class of compound upon which organic chemistry is based. For Ge the +2 state has some stable compounds, such as GeI_2, although the +4 state is predominant. Tin forms compounds in the +4 and +2 oxidation states, both of which are stable, although tin(+2) is a reducing agent, whereas for lead the +4 state is generally oxidizing and the +2 state is stable. Some lead compounds such as PbO_2 are powerful oxidizing agents.

The Group 14 elements have the ns^2np^2 electronic configuration, and so we might expect all these elements to form divalent compounds, forming two bonds using the two p-electrons. Promotion of an electron from ns^2 into the empty np orbital requires energy, but the atom can now form four bonds. More energy is produced by bond formation for carbon and

This trend (the **inert pair effect**) in oxidation state stabilities parallels those for the elements of Group 13 (oxidation states +1 and +3) and Group 15 (oxidation states +3 and +5).

(to a lesser extent) for silicon (which forms short, strong covalent bonds), which offsets the cost of promoting an ns electron. The tetravalent state is very stable for these two elements. For the heavy elements (Ge, Sn, Pb), bond strengths decrease, so the formation of four bonds cannot provide enough energy to stablize the +4 oxidation state. Thus, for lead (and, to a lesser extent, tin) the +2 state is the most stable.

A very important feature of Group 14 element chemistry is catenation; an enormous number of carbon compounds containing C–C bonds are known. Catenation becomes less important going down Group 14, owing to the decreased bond energies and the stabilization of the +2 oxidation state.

Catenation is the formation of rings and chains by atoms of the same element. After carbon, sulfur forms the next largest number of catenated compounds.

Worked Problem 6.1

Q What factors contribute to the stability of strong C–C bonds?

A Carbon has an intermediate electronegativity (2.5), so it is very willing to participate in homonuclear bond formation, by sharing electrons [elements with very low electronegativities (metals) tend to form cations and elements with very high electronegativities (F, O, Cl) tend to form anions]. Since carbon is small, overlap of orbitals in covalent bond formation is very effective, so C–C covalent bonds are very strong.

Average bond energies (kJ mol^{-1}):

C–C	+346
Si–Si	+226
N–N	+158
S–S	+266

6.2 The Elements

6.2.1 Carbon

Important sources of carbon are coal and crude oil; there are also substantial deposits of natural gas, which is mainly methane, CH_4. Carbon occurs as several allotropes, which have very different structures and properties. Until recently, only two *crystalline* forms of carbon – diamond and graphite – were known [many *amorphous* (non-crystalline) forms of carbon are familiar, *e.g.* soot].

In graphite, the carbon atoms form planar sheets of fused six-membered rings (Figure 6.1). Each carbon atom uses sp^2 hybrids to bond to other carbons within the sheet, and its remaining p$_z$ orbital forms an extensive delocalized π-system over the sheet. Attraction *between* adjacent sheets is weak, and so the graphite layers slide over each other very easily. The graphite structure is unique to carbon among the Group 14 elements, because the small p$_z$ orbitals on C can overlap effectively to form the delocalized π-system.

Allotropes are different structural forms of the same element, in which the chemical bonding is different. Another example is the ozone (O_3) and dioxygen (O_2) forms of oxygen.
Polymorphs are different crystalline forms of the same element (or compound), *e.g.* rhombic and monoclinic forms of S_8 (see Section 8.2.2).

Boron nitride (BN) is isoelectronic with carbon, and also adopts similar structures; see Section 5.8.2.

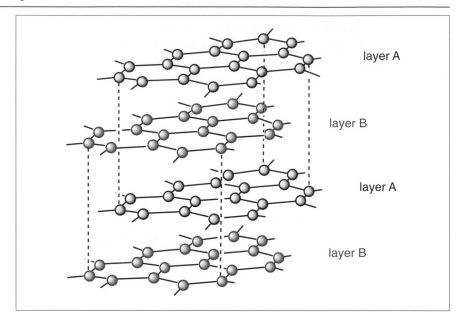

Figure 6.1 The structure of graphite, showing the superposition of alternating layers

In contrast, **diamond** has an infinite three-dimensional network structure built up from tetrahedral carbon atoms forming strong covalent bonds with each other (Figure 6.2). Diamond is therefore one of the hardest substances known.

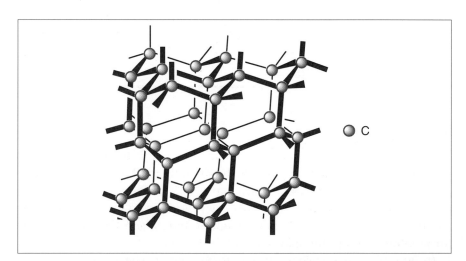

Figure 6.2 The three-dimensional network structure of diamond

Fullerenes and nanotubes are made by passing an electrical arc between two graphite electrodes under a helium atmosphere.

In recent years there has been an explosion of research into new *molecular* allotropes of carbon called **fullerenes**, typified by the first compound discovered, C_{60}, shown in Figure 6.3.[1] There is an extensive series of such materials, also termed Buckminsterfullerenes or 'buckyballs' after the architect R. Buckminster Fuller who designed geodesic domes. There are also so-called carbon nanotubes, 'buckytubes', which consist of a graphite-like tube, capped at both ends by a fullerene hemisphere.

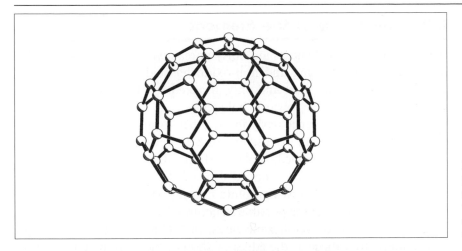

Figure 6.3 The structure of Buckminsterfullerene, C_{60}. The structure of the molecule, built from pentagons and hexagons of carbon atoms, is the same as a soccer ball

Worked Problem 6.2

Q The melting points of diamond and graphite are greater than 3550 °C, but C_{60} sublimes between 450 °C and 500 °C. Explain this observation.

A C_{60} is a *molecular* substance, so to convert it from the solid state to the gaseous state the relatively weak *intermolecular* forces must be broken (the intramolecular bonds within the C_{60} molecule are strong). In contrast, diamond and graphite have polymeric structures built up from strong C–C bonds. To melt diamond and graphite, some of these C–C bonds must be broken, which requires a lot of energy.

6.2.2 Silicon and Germanium

Silicon is the second most abundant element (after oxygen) in the Earth's crust, where it occurs as SiO_2 and silicate minerals. Germanium, the least common of the Group 14 elements, occurs in some zinc and silver ores, and also in flue dusts and the ashes of certain types of coal. Silicon and germanium adopt diamond-like structures (Figure 6.2).

6.2.3 Tin and Lead

Tin mainly occurs naturally as the mineral *cassiterite*, SnO_2, from which metallic tin is obtained by reduction.

The main ore of lead is the mineral *galena*, PbS, which is roasted in air to form lead oxide, which is then reduced to the metal using carbon monoxide in a blast furnace. Lead is a typical heavy metal.

6.3 Chemistry of the Elements

All of the Group 14 elements are fairly unreactive. Tin and lead dissolve in a number of oxidizing and non-oxidizing acids, though silicon and germanium are rather inert and only attacked by HF. All of the Group 14 elements are reactive towards halogens.

While diamond is particularly inert, other forms of carbon are more reactive. Graphite has an extensive chemistry and can form two types of products with either buckled or planar sheets, depending on whether the delocalized π-system is, or is not, disrupted. Thus, reaction of graphite with fluorine at 400 °C gives graphite fluoride, $(CF_x)_n$, which is colourless when $x = 1$. One of the possible structures for $(CF)_n$, shown in Figure 6.4, consists of fused cyclohexane rings; each C atom bears a single fluorine, apart from those at the edges of sheets which will be CF_2 groups. The fluorination prises apart the graphite layers, which become non-conducting, since delocalized electrons are no longer present.

Other graphite compounds, with planar graphite sheets, are inclusion compounds, formed when graphite is reacted with a reducing agent, typically an alkali metal, *e.g.* for KC_8 in equation 6.1.[2]

$$8C(\text{graphite}) + K_{(g)} \rightarrow KC_{8(s)} \qquad (6.1)$$

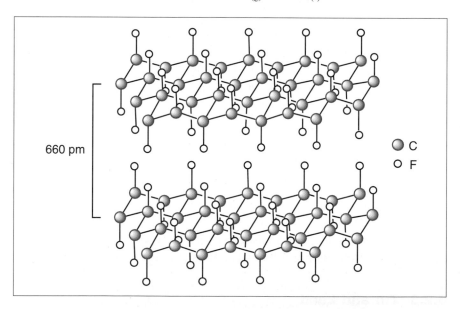

660 pm

○ C
○ F

Figure 6.4 The proposed structure of graphite fluoride, $(CF)_n$

In this compound (and related materials) the interlayer separation increases to accommodate the alkali metal atom, which forms a C_6–metal–C_6 sandwich. In KC_8 there are alternating layers of C and K (Figure 6.5); in materials with less potassium (*e.g.* KC_{24}), K layers are inserted into alternate carbon layers.

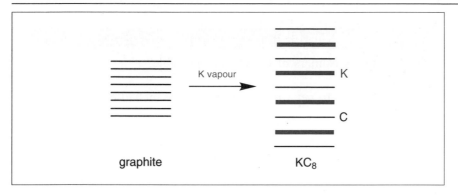

Figure 6.5 Formation and layered structure of KC_8 with planar graphite sheets

6.4 Group 14 Hydrides and Organometallic Compounds

Methane, CH_4, is the simplest hydride of carbon; the corresponding hydrides MH_4 of the other Group 14 elements are less stable owing to the size mismatch between the Group 14 atom and the very small H atom. Silane, germane and stannane are all known, but plumbane is very unstable and no chemistry has been determined. General methods for the synthesis of hydrides, including organo derivatives which are generally more stable than the simple hydrides, use hydrides such as $Li[AlH_4]$, as shown in equations 6.2–6.4.

Nomenclature
CH_4 Methane
SiH_4 Silane
GeH_4 Germane
SnH_4 Stannane
PbH_4 Plumbane

$$SiCl_4 + LiAlH_4 \rightarrow SiH_4 + LiAlCl_4 \tag{6.2}$$

$$GeO_2 + LiAlH_4 \rightarrow GeH_4 + LiAlO_2 \tag{6.3}$$

$$4PhSiCl_3 + 3LiAlH_4 \rightarrow 4PhSiH_3 + 3LiAlCl_4 \tag{6.4}$$

The chemistry of longer-chain carbon hydrides and derivatives thereof is covered by the subdiscipline of organic chemistry. We will say no more about these important compounds other than to stress again the propensity of carbon to form catenated compounds, with very strong bonds between two or more C atoms, which provides the diversity and stability of organic compounds. The stability of long-chain compounds markedly decreases for the heavier Group 14 elements. Catenation up to $Si_{10}H_{22}$ is established for silicon and to Ge_9H_{20} for germanium, but for tin, only Sn_2H_6 is known. As with hydrocarbons, different chain-branched isomers can often be formed for silanes and germanes.

Hydrocarbons are thermodynamically unstable in air; suitable activation (such as a spark) readily ignites hydrocarbon–air mixtures, *e.g.* octane (a fuel):
$2C_8H_{18} + 25O_2 \rightarrow 16CO_2 + 18H_2O + energy$

Worked Problem 6.3

Q How many isomers of Ge_5H_{12} are possible?

A It is possible to get straight-chain and branched-chain isomers:

Hydrogen atoms not shown

The first example of a compound with a Pb–Pb double bond

A wide range of alkyl and aryl compounds of Si, Ge, Sn and Pb have been prepared, *e.g.* Me_3SnCl and Et_4Pb. The latter has historically been used as an anti-knock additive in petrol.

For carbon, compounds containing double (C=C, C=O, C=N, *etc.*) and triple (*e.g.* C≡C, C≡N, C≡P) bonds are well known, but analogous compounds of the heavier elements are rare and require bulky organic groups on the multiple bond in order to stabilize them; compounds containing Si=Si, Ge=Ge, Sn=Sn and Pb=Pb bonds are all known,[3] the latter only recently.[4] The instability of multiple bonds between heavy p-block elements is largely due to poor orbital overlap.

6.5   Group 14 Halides

6.5.1 Tetrahalides, EX_4

All of the Group 14 tetrahalides exist, except for PbI_4 where the lead(IV) centre is too strong an oxidizing agent to co-exist with the reducing iodide; $PbBr_4$ is thermally very unstable, for the same reasons. The tetrahalides can be made by direct combination of the elements, sometimes via the dihalides (equations 6.5 and 6.6) or by treatment of the oxide with HX (equation 6.7).

$$Si + 2Cl_2 \rightarrow SiCl_4 \qquad (6.5)$$

$$Sn + 2I_2 \rightarrow SnI_2 \rightarrow SnI_4 \qquad (6.6)$$

$$GeO_2 + 4HBr \rightarrow GeBr_4 + 2H_2O \qquad (6.7)$$

Most EX_4 compounds are tetrahedral molecular species, and are quite volatile. However, the fluorides of tin and lead (SnF_4 and PbF_4) are solids with polymeric structures composed of fluoride-bridged MF_6 octahedra, with a significant ionic contribution to the bonding. This trend matches other p-block groups, such as Group 13 (where BF_3 is molecular but AlF_3 is polymeric) and Group 15 (where NF_3 and PF_3 are molecular but BiF_3 is ionic). In forming such polymeric compounds the larger Group 14 elements increase their coordination number. Silicon, germanium, tin and lead tetrahalides (EX_4) form coordination complexes with electron-pair donors (D), $EX_4.D$ (with trigonal bipyramidal geometry) and $EX_4.2D$ (with octahedral geometry). The donors may be neutral ligands, such as ethers, or anionic ligands such as halides. Thus, addition of Cl^- to $SnCl_4$ can give $SnCl_5^-$ and $SnCl_6^{2-}$.

The tetrahalides of Si, Ge, Sn and Pb (with the exception of SiF_4) hydrolyse rapidly to the hydrated dioxide (equation 6.8).

$$SiCl_4 + excess\ H_2O \rightarrow SiO_2.nH_2O + 4HCl \qquad (6.8)$$

Carbon tetrahalides are **kinetically stable** towards hydrolysis, because carbon (unlike the heavier group elements) is not able to increase its coordination number. For silicon and the heavier elements, an incoming water molecule can readily add to the SiX_4 group to form a five-coordinate intermediate which can then lose HCl.

6.5.2 Dihalides, EX_2

The +2 oxidation state is the important stable state for lead and it is also quite stable for tin, which forms all four dihalides by reaction of the metal with anhydrous HX (equation 6.9).

$$Sn + 2HX \rightarrow SnX_2 + H_2 \qquad (6.9)$$

$SnCl_2$ in the gas phase has a bent structure (since it is a three-electron pair molecule, with two Sn–Cl bonds and a lone pair on Sn). In the solid state it forms a chain polymer through the formation of chloride bridges.

Germanium(II) halides are known for all halides, formed by the reduction of GeX_4 with elemental Ge (equation 6.10), which is an example of **comproportionation**. GeI_2 disproportionates into GeI_4 and Ge on heating.

$$GeX_4 + Ge \rightarrow 2GeX_2 \qquad (6.10)$$

Adduct formation by addition of X^- to SnX_4

SF$_6$ is similarly stable towards hydrolysis; see Section 8.4.2.

The C–Cl bond itself shows no special stability towards hydrolysis; ethanoyl chloride (acetyl chloride, MeCOCl) very rapidly hydrolyses to ethanoic acid (acetic acid, $MeCO_2H$) because the water can readily attack the *three-coordinate* carbon atom.

Gas phase

Solid state

Comproportionation is the opposite of disproportionation (page 62).

6.5.3 Halides with E–E Bonds: Catenated Halides

Disproportionation is where some atoms of the same element are oxidized and others reduced in the same reaction.

For carbon, just as there is a diverse range of hydrocarbons known, there is a similarly extensive range of halocarbons; perhaps the best example is poly(tetrafluoroethene), PTFE, an extremely stable polymer finding numerous applications (see Section 9.2.1).

For silicon, a large number of higher halides are known, containing chains of Si atoms, analogous to the polysilanes described in Section 6.4. Germanium, tin and lead form few analogues of the silicon compounds, because of the lower stability of the E–E bonds going down the group, and the increased stability of the divalent halides (equation 6.11).

$$Ge_2Br_6 \xrightarrow{\text{heat}} GeBr_2 + GeBr_4 \qquad (6.11)$$

This reaction, when reversed, forms a convenient synthesis for compounds such as Ge_2Br_6 and Ge_2Cl_6.

6.6 Carbides and Silicides

When carbon or silicon is heated with many elements (especially metals), carbides and silicides are formed. There are many different types of carbides, which are classified by their reaction (or lack of it) with water, and the products formed therefrom.

Compare the formation of interstitial hydrides (Section 2.4.3).

Transition metals typically form interstitial carbide compounds, in which the individual carbon atoms occupy holes in close-packed metallic lattices; these materials are also very hard. Some interstitial carbides hydrolyse to give hydrogen and hydrocarbons, but others such as tungsten carbide (WC) are very hard, inert materials.

The carbides of reactive metals tend to be ionic in nature; some carbides (such as Na_4C, Be_2C) formally contain C^{4-} ions and give methane (CH_4) on hydrolysis, while others contain the acetylide ion (C_2^{2-}) and give ethyne (acetylene) (equation 6.12). The best-known example of this type is calcium carbide, CaC_2, which has an elongated sodium chloride-type structure (see Figure 3.4) with the C_2^{2-} ions arranged in a parallel fashion (Figure 6.6).

$$CaC_2 + 2H_2O \rightarrow C_2H_2 + Ca(OH)_2 \qquad (6.12)$$

Other carbides give longer-chain hydrocarbons on hydrolysis; Li_4C_3 gives propyne ($MeC{\equiv}CH$) (equation 6.13).

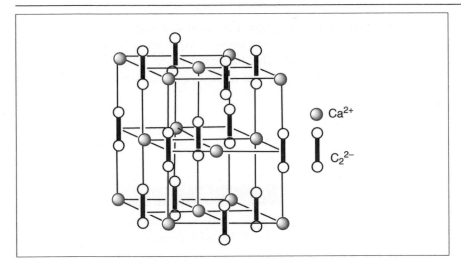

Ca^{2+}

C_2^{2-}

Figure 6.6 The structure of calcium carbide, CaC_2

$$Li_4C_3 + 4H_2O \rightarrow MeC{\equiv}CH + 4LiOH \qquad (6.13)$$

6.7 Oxides

The oxides of carbon, being gases, are quite different from those of the other Group 14 elements. The structural difference results from the presence of strong pπ–pπ bonding between carbon and oxygen; carbon oxides (and anions such as CO_3^{2-}) are discrete (non-polymeric) species. In contrast, for silicon the Si=O double bond is unstable with respect to two Si–O single bonds, and so silicon oxides and many oxyanions have infinite, covalent network structures of Si–O bonds (see Chapter 12).

There are some similarities in the preparations, if not in the structures, of the Group 14 oxides; heating the element in oxygen forms the dioxides, EO_2.

Group 14 element dioxides:
CO_2 linear molecule, weak acid
SiO_2 covalent polymer, weak acid
GeO_2 amphoteric
SnO_2 amphoteric
PbO_2 inert to acids and bases; oxidizing

6.7.1 Carbon Oxides

Carbon forms three main oxides, CO, CO_2 and C_3O_2.

Carbon monoxide (C≡O) is obtained by the dehydration of methanoic acid (formic acid) using concentrated H_2SO_4 (equation 6.14).

$$HCO_2H - H_2O \rightarrow CO \qquad (6.14)$$

CO could, therefore, be viewed as the anhydride of methanoic acid, but it is not a true anhydride because it is poorly soluble in water and does not react with it. It does, however, react with concentrated hydroxide solutions on heating, to give the methanoate (formate) anion (equation 6.15).

$$CO + OH^- \rightarrow HCO_2^- \qquad (6.15)$$

	CO	N_2
M.p. (°C)	–205	–210
B.p. (°C)	–190	–196

CO is isoelectronic with nitrogen (N_2) and has similar physical properties. However, CO, which is very poisonous, is much more reactive than nitrogen. It combines with the halogens (except iodine) directly, for example with chlorine to give the highly poisonous **phosgene** (equation 6.16).

$$CO + Cl_2 \rightarrow COCl_2 \text{ (phosgene)} \qquad (6.16)$$

Worked Problem 6.4

Q CO is a strong reducing agent and can reduce most metals from their oxides to the metal itself. Write balanced equations for the reduction of Fe_2O_3 to Fe, and SnO_2 to Sn, using CO.

A The by-product is CO_2:

$$Fe_2O_3 + 3CO \rightarrow 2Fe + 3CO_2$$
$$SnO_2 + CO \rightarrow Sn + 2CO_2$$

A fuller account of metal carbonyl compounds, their chemistry, structures and bonding is to be found in the Tutorial Chemistry Texts book *Organotransition Metal Chemistry*.

Carbon monoxide forms numerous complexes with metals in low and intermediate oxidation states,[5] where it bonds through the carbon alone. Some examples are given in Figure 6.7.

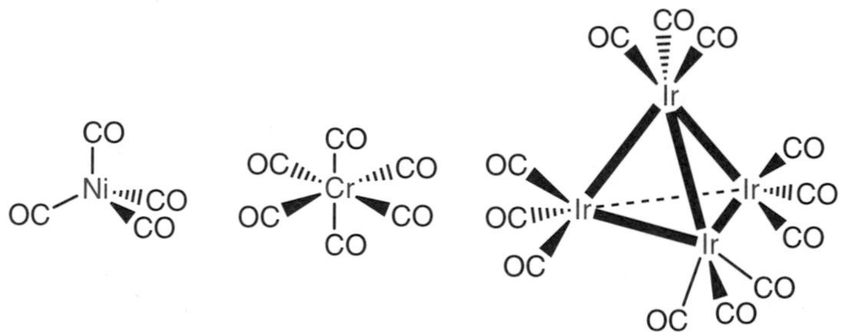

Figure 6.7 Some metal carbonyls

Carbon dioxide (CO_2), the most stable carbon oxide, is produced on an enormous scale industrially by the combustion of coal, oil and natural gas. Alternatively, it may be generated by the thermal decomposition of limestone (calcium carbonate) (equation 6.17).

$$CaCO_3 \rightarrow CaO + CO_2 \qquad (6.17)$$

Carbon dioxide is soluble in water, mostly as dissolved CO_2 molecules, but a small amount of CO_2 is hydrated to give carbonic acid, H_2CO_3.

This is a weak acid, only partly dissociated to **hydrogencarbonate** (bicarbonate, HCO_3^-) or **carbonate** (CO_3^{2-}) ions. The carbonate ion is obtained when CO_2 is passed through an aqueous hydroxide solution (*e.g.* NaOH) (equation 6.18). Passage of more CO_2 gives the hydrogencarbonate ion. When carbonates of Group 1 or 2 metals are treated with CO_2 (and water), bicarbonates are formed (equation 6.19).

$$2OH^- + CO_2 \rightarrow CO_3^{2-} + H_2O \xrightarrow{\text{more } CO_2} 2HCO_3^- \qquad (6.18)$$

$$CaCO_{3(s)} + CO_{2(g)} + H_2O_{(l)} \rightleftharpoons Ca(HCO_3)_{2(aq)} \qquad (6.19)$$

Most metal carbonates are insoluble in water, except for ammonium carbonate and the alkali metal carbonates, and so can be prepared by adding a sodium carbonate solution to the aqueous metal ion, *e.g.* equation 6.20. However, this cannot be used for $Al_2(CO_3)_3$ since the aqueous Al^{3+} ion is highly acidic, simply decomposing the carbonate ion to CO_2.

$$Cd^{2+}_{(aq)} + CO_3^{2-}_{(aq)} \rightarrow CdCO_{3(s)} \qquad (6.20)$$

The carbonate ion is planar, with equal C–O bond distances.

Carbon suboxide (C_3O_2) is the third oxide, made by dehydrating propanedioic acid (malonic acid) with P_4O_{10} (equation 6.21).

$$CH_2(CO_2H)_2 - 2H_2O \rightarrow O{=}C{=}C{=}C{=}O \qquad (6.21)$$

The C_3O_2 molecule is linear, with $p\pi$–$p\pi$ bonding similar to that in carbon dioxide.

6.7.2 Silicon, Germanium, Tin and Lead Oxides

The structural chemistry of SiO_2 is extremely complex; the stable form of SiO_2 under ambient conditions is *quartz*, and other high-temperature (*tridymite* and *cristobalite*) modifications are known. The various SiO_2 structures differ in the way in which SiO_4 tetrahedra are linked together. A vast number of polymeric silicates occur naturally; a brief survey of their structural features is presented in Chapter 12.

The dioxides GeO_2, SnO_2 and PbO_2 have considerable ionic character and adopt the *rutile* (TiO_2) structure (Figure 6.8), with a six-coordinate metal atom and three-coordinate oxygen. The monoxides of tin (SnO) and lead (PbO, *litharge*) are also well known. Addition of hydroxide to aqueous solutions of Ge^{2+} (in the absence of air), Sn^{2+} or Pb^{2+} precipitates the hydrated oxides (equation 6.22).

$$M^{2+}_{(aq)} + 2OH^-_{(aq)} \rightarrow MO.nH_2O_{(s)} \qquad (6.22)$$

Unlike $NaHCO_3$, Group 2 hydrogencarbonates cannot be obtained as solids.

This equilibrium is very important in geochemistry, for example in the deposition and redissolution of limestone rock (principally $CaCO_3$) during weathering.

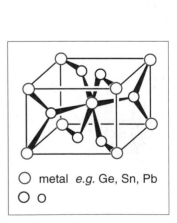

○ metal *e.g.* Ge, Sn, Pb
○ O

Figure 6.8 The solid-state rutile (TiO_2) structure adopted by MO_2 (M = Ge, Sn or Pb)

These oxides are amphoteric, and dissolve in both acids and bases. The hydrated Sn^{2+} and Pb^{2+} ions undergo substantial hydrolysis.

Worked Problem 6.5

Q Lead forms the mixed oxidation state oxide Pb_3O_4 ('red lead'). Calculate the theoretical yield of lead(IV) ethanoate, $Pb(OAc)_4$, by reaction of 1 gram of Pb_3O_4 with excess pure ethanoic acid (AcOH) containing some Ac_2O to remove water [note: PbO and PbO_2 react with AcOH to form $Pb(OAc)_2$ and $Pb(OAc)_4$, respectively].

A The composition of Pb_3O_4 is $(PbO)_2(PbO_2)$, so the reaction with AcOH is:

$$2PbO + 4AcOH \rightarrow 2Pb(OAc)_2 + 2H_2O$$
$$PbO_2 + 4AcOH \rightarrow Pb(OAc)_4 + 2H_2O$$

Thus: $Pb_3O_4 + 8AcOH \rightarrow 2Pb(OAc)_2 + Pb(OAc)_4 + 8H_2O$ (removed by Ac_2O), so 1 gram (0.0015 mole) of Pb_3O_4 will give 0.0015 mole (0.66 grams) of $Pb(OAc)_4$.

Removal of water: $Ac_2O + H_2O \rightarrow 2AcOH$

Worked Problem 6.6

Q Why is the Sn^{2+} ion more extensively hydrolysed in aqueous solution than Pb^{2+}?

A The Sn^{2+} ion has a higher charge density than Pb^{2+} owing to its smaller ionic radius (Sn^{2+} 93 pm, Pb^{2+} 119 pm), so it will polarize the water molecules more strongly, resulting in increased loss of H^+ and formation of complexes with coordinated hydroxide ions.

6.8   Sulfides, Selenides and Tellurides

Compare metal–CO complexes (Section 6.7.1).

Carbon disulfide (CS_2) is a molecular substance, analogous to CO_2. The species CS (the sulfur analogue of carbon monoxide) is unstable as a free molecule, but can be stabilized by coordination to a metal. Several transition metal–CS complexes are known, for example $[RhCl(CS)(PPh_3)_2]$. The compounds ME and ME_2 (E = S, Se, Te) are known for Ge, Sn and Pb.

6.9   Polyatomic Anions of the Group 14 Elements

When alloys of heavy Group 14 elements (Ge, Sn and Pb) and alkali metals are dissolved in liquid ammonia or ethane-1,2-diamine (ethyl-enediamine, $NH_2CH_2CH_2NH_2$), polyatomic anions are produced; these can be crystallized if a donor ligand, such as a crown ether (see Section 3.9), is added to the solution to complex the alkali metal cation. A wide variety of such polyanions have been prepared, such as Sn_4^{2-}, Ge_4^{2-} (which are tetrahedral), Sn_5^{2-}, Pb_5^{2-} (trigonal bipyramidal) and Sn_9^{4-} (a capped square antiprism, Figure 6.9).

Wade's rules (Section 5.6.1 and Box 5.1) can be used to rationalize the structures of many of these clusters, with one additional step. Each (non-hydrogen bearing) atom generally retains a lone pair of electrons, so these must be subtracted from the total electron count when determining the number of skeletal electron pairs (SEP).

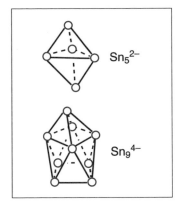

Figure 6.9 Structures of some tin polyanions

Worked Problem 6.7

Q Using Wade's rules, confirm that the Sn_5^{2-} ion has a trigonal bipyramidal shape.

A By reference to Box 5.1, each Sn atom has 4 valence electrons, but retains 2 in a lone pair. Hence total electron count is (5×2), plus 2 for the negative charge = 12 electrons, or 6 pairs. Thus, 6 SEP gives a five-vertex polyhedron, *i.e.* a trigonal bipyramid. There are 5 vertices and 5 Sn atoms, so the cluster is of the *closo* type.

Summary of Key Points

1. The *abundance and stability* of compounds containing C–C bonds and *the increasing stabilization of the +2 oxidation state* for the heavier elements are two factors which dominate the chemistry of the Group 14 elements.

2. *pπ bonding* is very important for carbon.

3. For silicon, the *Si–O single bond is very strong*, resulting in a wide range of polymeric oxides.

Problems

6.1. Determine the oxidation state of the Group 14 element in the following compounds or ions: (a) COF_2; (b) $SnCl_3^-$; (c) $[Pb_6(OH)_8]^{4+}$; (d) $Ph_3Pb-PbPh_3$.

6.2. Identify the element X in each of the following:
(a) The oxide XO_2 has a high melting point, and is very abundant in Nature.
(b) X forms three oxides, XO, XO_2 and X_3O_2.
(c) X forms compounds mainly in the +2 oxidation state, though some compounds in the +4 state do exist.
(d) X occurs as several allotropes, including a molecular one.

6.3. Compounds of carbon typically show chemical and physical properties different to those of silicon. Summarize some of the differences.

6.4. Why do the enthalpies of formation (kJ mol^{-1}) of the Group 14 hydrides EH_4 become more positive in the order: CH_4 (−74), SiH_4 (+34), GeH_4 (+91), SnH_4 (+163)?

6.5. Explain what is meant by: (a) catenation; (b) allotropes; (c) an endohedral fullerene complex; (d) a disproportionation reaction.

6.6. Predict the outcome of the following reactions, and write balanced equations:
(a) Sn + excess $I_2 \rightarrow$
(b) $Be_2C + H_2O \rightarrow$
(c) $CCl_4 + H_2O \rightarrow$
(d) $Et_2SiCl_2 + LiAlH_4 \rightarrow$

6.7. Draw the different possible isomers of the silane Si_6H_{14}.

References

1. H. W. Kroto, *Chem. Br.*, September 1996, 32; H. W. Kroto, *Angew. Chem., Int. Ed. Engl.*, 1992, **31**, 111.
2. R. Csuk, B. I. Glänzer and A. Fürstner, *Adv. Organomet. Chem.*, 1988, **28**, 85.
3. T. Tsumuraya, S. A. Batcheller and S. Masamune, *Angew. Chem., Int. Ed. Engl.*, 1991, **30**, 902.
4. M. Stürmann, W. Saak, H. Marsmann and M. Weidenbruch, *Angew. Chem., Int. Ed. Engl.*, 1999, **38**, 187.
5. E. W. Abel, *Educ. Chem.*, 1992, 46.

Further Reading

Buckminsterfullerenes, ed. W. E. Billups and M. A. Ciufolini, VCH, Weinheim, 1993.

7

The Group 15 (Pnictogen) Elements: Nitrogen, Phosphorus, Arsenic, Antimony and Bismuth

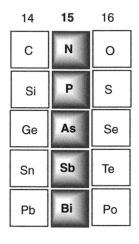

Aims

By the end of this chapter you should understand:

- The diversity of oxides, sulfides, halides and hydrides of the Group 15 elements
- The difference in physical and chemical properties of the elements

7.1 Introduction and Oxidation State Survey

The Group 15 elements, in the centre of the p-block, display a wide range of chemical and physical properties in both their elements and chemical compounds. The lightest element (nitrogen) is a typical non-metal, while the heaviest (bismuth) has characteristic properties of a main group metal; there is a gradual decrease in the first (and also second and third) ionization energies on progressing from nitrogen to bismuth (Figure 7.1). Oxidation states of +3 and +5 occur for all elements, with the +3 state being most stable for bismuth as a result of the inert pair effect (see Section 5.1). The +5 oxidation state is very stable for phosphorus, but acts as an oxidizing agent with nitrogen, arsenic and antimony. However, arsenic, being in the row of non-metals following the 3d elements (with a filled 3d inner shell which is relatively poorly shielding) is rather more difficult to oxidize to the +5 oxidation state than might be expected, because the s-electrons are tightly held and so are less likely to be involved in bonding. The −3 oxidation state occurs in the hydrides such as NH_3, and in anions such as P^{3-} (in Na_3P).

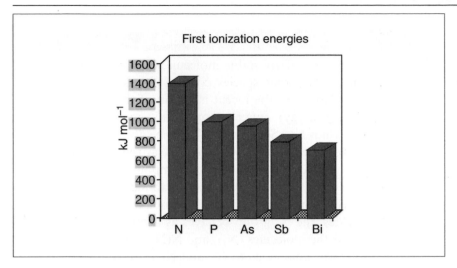

Figure 7.1 First ionization energies for the Group 15 elements

7.2   The Elements

Nitrogen is the most abundant gas in the atmosphere, constituting 78.1% by volume (Figure 7.2). Elemental nitrogen occurs exclusively as the triply bonded dinitrogen molecule, N_2, and is generally considered to be extremely unreactive. Other high-energy allotropes of nitrogen are also possible (Box 7.1).

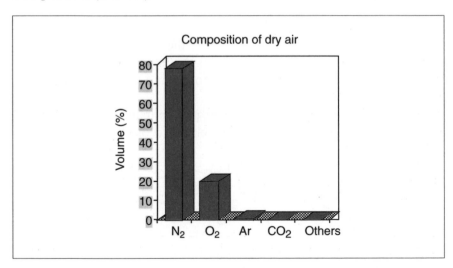

Figure 7.2 The composition of dry air

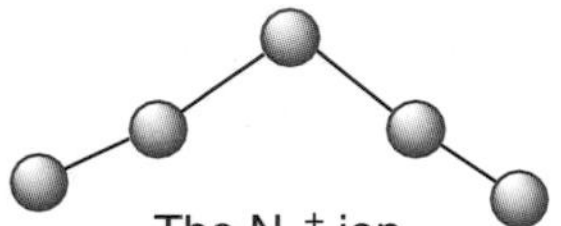

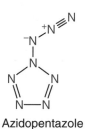

The N_5^+ ion

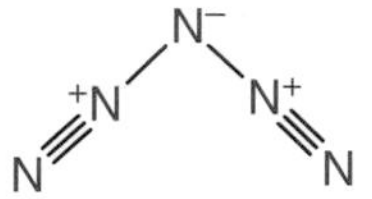

Simplified Lewis diagram for N_5^+

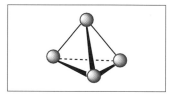

Azidopentazole

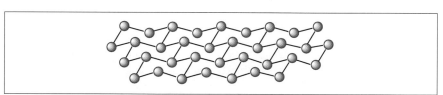

Figure 7.3 The structure of white phosphorus, P_4

Box 7.1 High-energy Nitrogen Species

Because N_2 is a particularly stable molecule, both thermodynamically and kinetically, other species comprised solely of nitrogen atoms are rare. However, the recent isolation of the N_5^+ ion[1] suggests that a (high-energy) *ionic* allotrope of nitrogen $N_5^+N_3^-$ (containing the azide ion N_3^-, Section 7.3.2) may be able to be isolated. Another N_8 allotrope predicted to have a relatively low energy is **azidopentazole**. There is interest in such compounds as high-energy-density materials, and many other homoatomic nitrogen species can be envisaged (though their synthesis has not yet been achieved). In covalent azides, the N_3 group behaves as a **pseudohalogen** (see Box 9.2), so that the molecules $(N_3)_2$ and $N(N_3)_3$ are also potential nitrogen allotropes, analogous to Cl_2 and NCl_3, respectively.

The principal source of **phosphorus** is *phosphate rock*, a complex calcium phosphate. Reduction of phosphate rock gives white phosphorus, with a tetrahedral structure (Figure 7.3) which persists in solid, liquid and gaseous phases. When P_4 vapour is heated above 800 °C, however, appreciable quantities of the triply bonded P_2 molecule are formed. Prolonged heating of white phosphorus in a sealed vessel results in the formation of **red phosphorus**, a relatively inert form of the element, with a poorly defined but complex structure.[2] However, the most stable allotrope is **black phosphorus**, formed by heating phosphorus at high pressure, which has several structures; the rhombohedral form has a layer structure shown in Figure 7.4.

Arsenic, **antimony** and **bismuth** occur predominantly as sulfides. The three elements form layered structures related to black phosphorus.

Figure 7.4 The structure of rhombohedral black phosphorus

Worked Problem 7.1

Q Single, double and triple bond strengths for N–N (and P–P) bonds are given in Table 7.1. Show that these are consistent with the occurrence of nitrogen as solely a triple-bonded diatomic molecule.

A The N≡N triple bond energy is *more than* three times that of the N–N single bond energy, so N–N single bonded allotropes are unstable relative to N_2. The N≡N bond energy is also greater than the sum of the N=N and N–N bond energies (in a –N=N–N=N– form), so this is also unfavourable relative to the N≡N bond.

Table 7.1 E–E single, double and triple bond energies for nitrogen and phosphorus

Bond	Bond energy (kJ mol^{-1})
N–N	+160
P–P	+200
N=N	+419
P=P	+310
N≡N	+945
P≡P	+490

7.3 Hydrides

7.3.1 Hydrides of the Type EH$_3$

All five Group 15 elements (E) form hydrides EH$_3$, though there is a marked and systematic trend in their physical and chemical properties. Thus, the boiling points increase on going down the group, with the exception of the anomalously high boiling point of NH_3, which is due to H⋯N–H hydrogen bonding (compare Section 2.6.1). The H–E–H bond angles also decrease going down the group; see Worked Problem 7.2. Owing to the decreasing E–H bond strengths (Table 7.2), thermal stabilities decrease down the group. The enthalpies of formation (Table 7.2) parallel the E–H bond energies for these compounds.

Table 7.2 Standard enthalpies of formation, $\Delta_f H°$, and average E–H bond energies for the Group 15 hydrides EH$_3$

	$\Delta_f H°$ (kJ mol^{-1})	E–H bond energy (kJ mol^{-1})
NH$_3$	−46.2	+391
PH$_3$	+9.3	+322
AsH$_3$	+172.2	+247
SbH$_3$	+142.8	+255

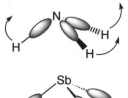

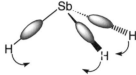

Nomenclature

	Traditional name	IUPAC[a] name
NH_3	ammonia	azane
PH_3	phosphine	phosphane
AsH_3	arsine	arsane
SbH_3	stibine	stibane
BiH_3	bismuthine	bismuthane

[a]International Union of Pure and Applied Chemistry

Worked Problem 7.2

Q Rationalize the trend in the H–E–H bond angles in the series of hydrides:

Molecule EH_3	H–E–H angle (°)
NH_3	107
PH_3	93.5
AsH_3	92
SbH_3	91.5

A All species have four valence pairs of electrons in the central E atom, corresponding to a tetrahedral arrangement. The H–E–H bond angles are expected to be <109.5° (the tetrahedral angle) owing to the effect of the lone pair. Distortions to the basic shape can be obtained by considering the electron distribution in the E–H bonds. The N–H bond is short (101.5 pm) and nitrogen is more electronegative than hydrogen, so the bonding pair will reside closer to the central N atom, occupying more space in its valence shell. These N–H bonding pairs will repel each other more, widening the bond angle.

Ammonia is manufactured on a huge scale by the chemical industry, by the reaction of nitrogen with hydrogen carried out over a catalyst (equation 7.1).

This synthesis is called the Haber Process.

$$N_2 + 3H_2 \xrightarrow[\text{catalyst}]{\text{high pressure and temperature,}} 2NH_3 \qquad (7.1)$$

Liquid ammonia (boiling point –33 °C) has comparisons with water as a solvent; H_3O^+ and OH^- are respectively an acid (a proton donor) and a base (a proton acceptor) in water, while in ammonia, NH_4^+ is an acid and NH_2^- is a base.

Worked Problem 7.3

Q Predict the outcome of the following reactions in liquid ammonia solvent:

(a) $Na_{(liq\ NH_3)} + NH_4Cl \rightarrow$

(b) $NH_4Cl + NaNH_2 \rightarrow$

A (a) NH_4^+ is an acid in liquid NH_3, so this is the reaction of metal + acid, which gives salt + hydrogen:

$$2Na_{(liq\ NH_3)} + 2NH_4Cl \rightarrow 2NaCl + 2NH_3 + H_2$$

(b) NH_2^- is a base in liquid NH_3, so this reaction is acid + base, which gives salt plus *ammonia*:

$$NH_4Cl + NaNH_2 \rightarrow NaCl + 2NH_3$$

Na in liquid NH_3: see Section 3.6.

Phosphine, PH_3, is made by reaction of P_4 with water (equation 7.2).

$$2P_4 + 12H_2O \rightarrow 5PH_3 + 3H_3PO_4 \qquad (7.2)$$

Phosphine is a highly toxic gas.

Ultrapure PH_3, required by the electronics industry, is manufactured by the thermal disproportionation of phosphorous acid, H_3PO_3 (equation 7.3).

$$4H_3PO_3 \rightarrow PH_3 + 3H_3PO_4 \qquad (7.3)$$

Phosphorous acid: Section 7.6.2.

The other hydrides, AsH_3, SbH_3 and BiH_3, can be synthesized by the reduction of the corresponding halide with a hydride source, such as $Li[AlH_4]$ (see Section 5.6.3); however, the yield of BiH_3 is low. The hydrolysis of phosphides, arsenides and antimonides of reactive metals such as calcium and aluminium can also generate phosphine, arsine or stibine (see also Section 2.4.2).

7.3.2 Other Hydrides

Of these, **hydrazine**, N_2H_4, is the most well-known compound, prepared by the oxidation of ammonia with chlorate(I) (hypochlorite) (equations 7.4 and 7.5).

$$NH_3 + ClO^- \rightarrow NH_2Cl + OH^- \qquad (7.4)$$

then

$$NH_2Cl + 2NH_3 \rightarrow N_2H_4 + NH_4Cl \qquad (7.5)$$

The structure of hydrazine consists of two NH_2 groups connected by an N–N bond, the molecule adopting the *gauche* conformation in the gas phase (Figure 7.5).

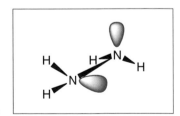

Figure 7.5 The structure of hydrazine, N_2H_4

Compounds E_2H_4 are known for E = N, P and As.

Worked Problem 7.4

Q Explain why the bond energy for the P–P single bond ($+200$ kJ mol^{-1}) is greater than both the N–N single bond ($+160$ kJ mol^{-1}) and the As–As single bond ($+175$ kJ mol^{-1}).

The same reason for a weak N–N bond also accounts for the relative weakness of related molecules containing O–O and F–F bonds (see Section 9.2).

A The general trend going down a group of the Periodic Table is that the E–E bond energies decrease owing to poorer overlap between spatially more diffuse orbitals on increasingly large atoms. This is shown by the low strength of the As–As bond (175 kJ mol^{-1}). Therefore, it might be expected that the N–N bond is the strongest. However, the lone pairs of electrons on adjacent N atoms repel each other (owing to the small atomic size), weakening the N–N bond.

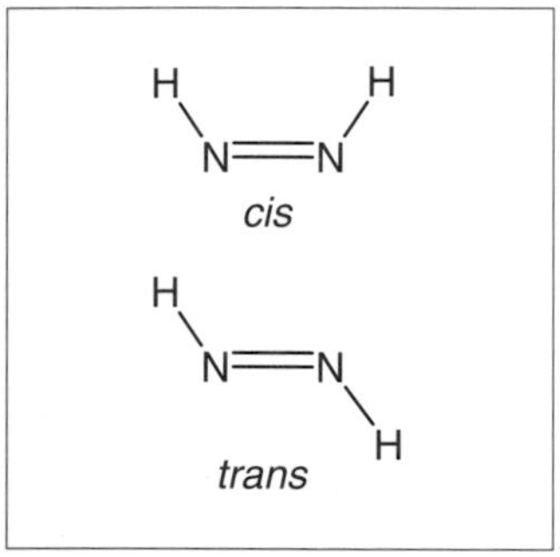

Figure 7.6 The *cis* and *trans* forms of diazene, N_2H_2

Diazenes of the type RN=NR (R = H or F) containing a nitrogen–nitrogen double bond are known, but N_2H_2 has only a transient existence. *Cis* and *trans* isomers are possible (Figure 7.6). The other elements show a decreased tendency to form this type of doubly bonded compound, owing to the decreased E=E bond strength on going down the group (see Table 7.1 for N and P), though a number of P=P and As=As containing compounds have recently been synthesized.

Another compound which can formally be considered as a nitrogen hydride is **hydrazoic acid**, HN_3, with the structure shown in Figure 7.7. The linear azide ion, N_3^-, can be formed by deprotonation of HN_3, and is symmetrical with two identical N–N bond lengths of 116 pm. The azide ion is an example of a **pseudohalogen** (see Box 9.2).

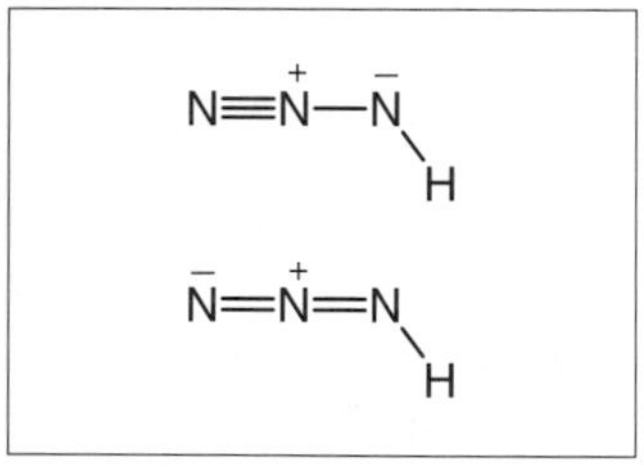

Figure 7.7 Two resonance forms of hydrazoic acid, HN_3

7.4   Oxides

All of the Group 15 elements form a wide range of oxides, particularly nitrogen and phosphorus.

7.4.1 Nitrogen Oxides

Oxides of nitrogen cover a range of oxidation states, from +1 to +5, and in all compounds there is significant N–O pπ–pπ bonding. Shapes of the main nitrogen oxides are shown in Figure 7.8.

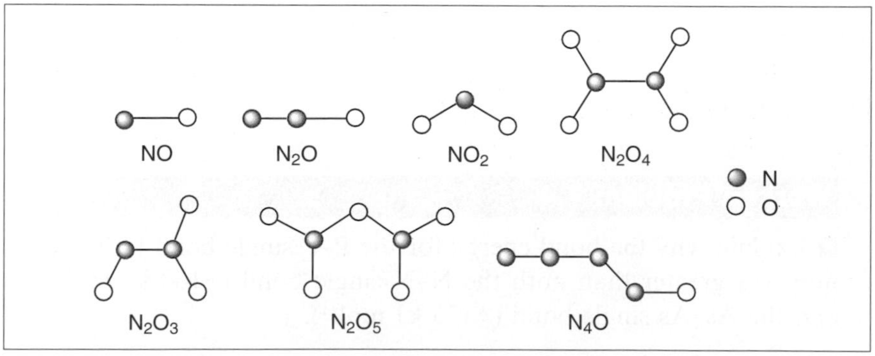

Figure 7.8 The shapes of nitrogen oxides

Dinitrogen oxide (nitrous oxide, N_2O), made by heating ammonium nitrate (equation 7.6), has a linear N–N–O structure.

$$NH_4NO_3 \rightarrow N_2O + 2H_2O \qquad (7.6)$$

Worked Problem 7.5

Q The enthalpy of formation of dinitrogen oxide:

$$2N_2 + O_2 \rightarrow 2N_2O$$

is +164 kJ mol^{-1}, yet the compound is relatively stable, and quite unreactive towards many reagents. Suggest a reason for this.

A With a positive enthalpy of formation, N_2O is *thermodynamically unstable* with respect to N_2 and O_2 (it therefore cannot be synthesized from the elements). However, N_2O is *kinetically stable*, and the decomposition at room temperature is extremely slow. NO is another nitrogen oxide which shows similar kinetic stability.

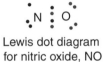

Lewis dot diagram
for nitric oxide, NO

Lewis dot diagram for
nitrosonium ion, NO$^+$

Odd-electron molecules formed by main group elements are relatively rare. Another example is ClO_2.

NO and N_2O are neutral oxides: solutions in water are neither acidic nor basic.

Nitrogen monoxide (nitric oxide, NO) is a reactive molecule containing one unpaired electron, but with little tendency to dimerize (to N_2O_2) except at low temperatures. NO reacts with halogens (X_2) to give **nitrosyl halides** of the type X–N=O, which have bent structures. Reaction of nitrosyl halides with a halide acceptor (*e.g.* $AlCl_3$) gives compounds containing the **nitrosonium cation**, NO$^+$ (equation 7.7). NO is easily oxidized by air to give brown NO_2.

$$O=N\text{–}Cl + AlCl_3 \rightarrow NO^+AlCl_4^- \qquad (7.7)$$

Worked Problem 7.6

Q Explain why the bond length in NO (115 pm) is longer than that in the nitrosonium (NO$^+$) ion (106 pm).

MO diagrams: Section 1.5

NO is a biologically important molecule, *e.g.* in the control of blood pressure.[3]

A We need to consider the molecular orbital (MO) diagram for NO and NO^+:

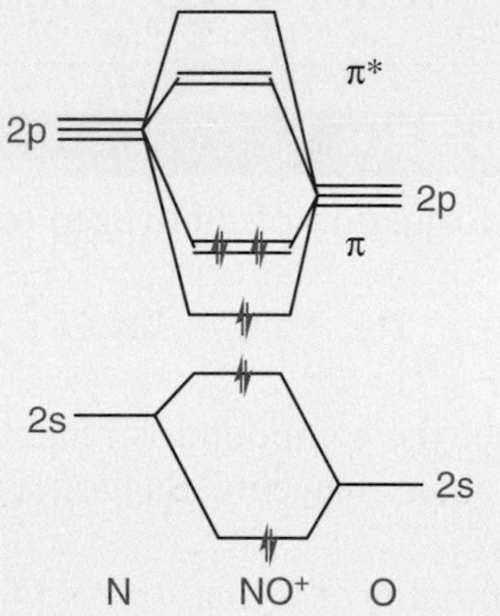

This is a heteronuclear diatomic molecule, and so the diagram (assuming no mixing of s- and p-orbitals) will be very similar to that of O_2 (see Section 1.5), but with some distortion due to the differing energies of the contributing N and O atomic orbitals. In NO there is a single electron in the π^* antibonding MO, giving a total bond order of 2.5. In NO^+ the electron in π^* is removed, and the net bond order will now be 3 (NO^+ is isoelectronic with N_2). Thus, NO^+ has a higher bond order than NO, and a shorter bond.

The formation of N_2O_3 and N_2O_4 occurs by pairing up odd electrons from $NO + NO_2$ and $NO_2 + NO_2$, respectively.

N_2O_3, NO_2, N_2O_4 and N_2O_5 are acidic oxides.

Other nitrogen oxides are N_2O_3, N_2O_4, N_2O_5 and N_4O, shown in Figure 7.8. N_2O_3 is formed from NO and NO_2 at low temperatures, where it is a blue solid or liquid, but it dissociates back to NO and NO_2 in the gas phase. N_2O_4 is colourless and diamagnetic when pure, but it readily dissociates upon warming to give the brown, paramagnetic NO_2 (equation 7.9). N_2O_5 contains covalent molecules in the gas phase, but crystallizes as nitronium nitrate, $NO_2^+NO_3^-$. It is usually formed by dehydrating nitric acid (equation 7.8), to which it is converted on reaction with water.

$$2HNO_3 - H_2O \rightarrow N_2O_5 \qquad (7.8)$$

Worked Problem 7.7

Q Predict the effect on the position of equilibrium in equation 7.9 of: (a) increasing pressure and (b) increasing temperature.

$$N_2O_{4(g)} \rightleftharpoons 2NO_{2(g)} \qquad \Delta H \text{ positive} \qquad (7.9)$$

A By Le Chatelier's principle, increased pressure will cause the equilibrium position to shift to the left, to favour N_2O_4, because there are two gaseous molecules on the right and one on the left. Since the reaction is endothermic, increased temperature will favour the forward reaction, producing more NO_2.

Le Chatelier's principle: for an equilibrium system, application of any disturbance will result in a shift in the position of equilibrium to try and eliminate the effect.

7.4.2 Phosphorus Oxides

A very wide range of phosphorus oxides are known,[4] the most important oxide being 'phosphorus pentoxide' [phosphorus(V) oxide, P_4O_{10}], formed by the combustion of phosphorus in an excess of oxygen (equation 7.10).

$$P_4 + 5O_2 \rightarrow P_4O_{10} \qquad (7.10)$$

It reacts violently with water, ultimately forming orthophosphoric acid on boiling (Section 7.6.2). Combustion of P_4 in a limited supply of oxygen gives the P(III) oxide P_4O_6. Related phosphorus oxides derived from the same basic P_4O_6 skeleton contain different numbers of terminal P=O groups; the structures of P_4O_6, P_4O_7 and P_4O_{10} are shown in Figure 7.9.

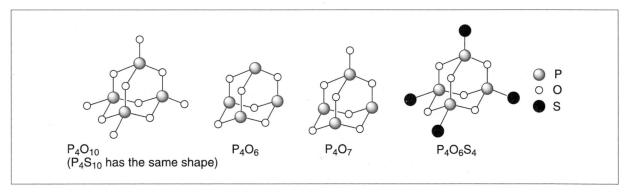

P_4O_{10}
(P_4S_{10} has the same shape)

P_4O_6

P_4O_7

$P_4O_6S_4$

○ P
○ O
● S

7.4.3 Arsenic, Antimony and Bismuth Oxides

Oxides of composition M_2O_5 are also formed for arsenic, antimony and bismuth, though the latter cannot be obtained pure. The structure of As_2O_5 is based on linked AsO_4 tetrahedra, while Sb_2O_5 is thought to contain linked MO_6 octahedra. Arsenic and antimony oxides E_2O_3 exist both as polymeric forms with edge-sharing EO_3 pyramids, and also in molecular E_4O_6 species, analogous to the phosphorus species P_4O_6 (Figure 7.9) Bi_2O_3 is polymeric with five-coordinate bismuth.

Figure 7.9 The structure of some important phosphorus oxides and sulfides

The heavier Group 15 elements tend to have higher coordination numbers.

7.5   Sulfides

There are few sulfides directly analogous to the wide range of molecular nitrogen oxides which are known.

The most well-known sulfur–nitrogen compound is the explosive S_4N_4, prepared by the reaction of SCl_2 with ammonia. The structure of the molecule is given in Figure 7.10; it contains delocalized π-bonding. Reaction of S_4N_4 with silver gives firstly S_2N_2 and then a polymeric substance $(SN)_x$ (Figure 7.10), which has very high electrical conductivity.

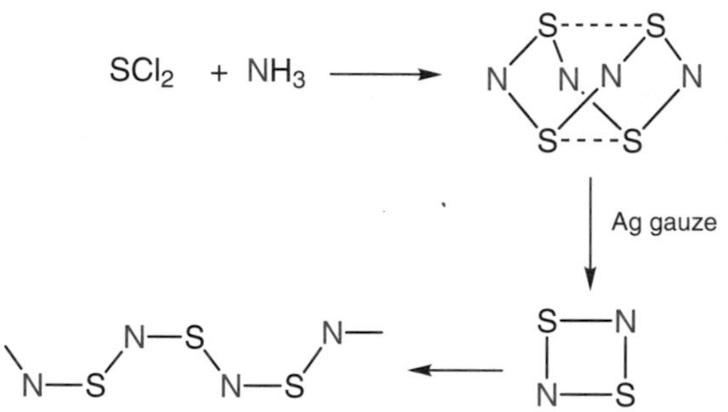

Figure 7.10 Some sulfur–nitrogen species

The most important sulfide of phosphorus is P_4S_{10}, which has a structure analogous to the oxide shown in Figure 7.9. Mixed oxide–sulfide compounds are also known, such as $P_4O_6S_4$, which has terminal P=S groups and bridging oxygens (Figure 7.9). A wide variety of lower oxidation state phosphorus sulfides exist and, as with the oxides, these can conceptually be derived from the tetrahedral P_4 molecule by addition of terminal P=S groups and insertion of S into P–P bonds; some of the compounds retain some P–P bonds, and are shown in Figure 7.11. Arsenic forms a number of sulfides. As_2S_3 exists as As_4S_6, with the same structure as P_4O_6 (Figure 7.9). There are also As_4S_3 (which occurs naturally as the mineral *realgar*), As_4S_5 and As_4S_{10}. The compound As_4S_4 exists in a number of forms, one of which is similar to that of S_4N_4 (Figure 7.10).

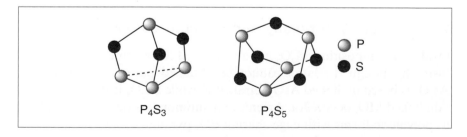

Figure 7.11 Some lower oxidation state phosphorus sulfides with P–P bonds

7.6   Oxyanions and Oxyacids

Oxyanions form an extremely important role in the chemistry of the Group 15 elements, in particular for nitrogen and phosphorus and to a slightly lesser extent for arsenic. On going down the group, to antimony and bismuth, oxyanions play a much smaller role in the chemistry of these elements, and they begin to show more cationic chemistry.

7.6.1 Oxyacids of Nitrogen

In all nitrogen oxyacids the nitrogen atom is either two or three coordinate (which contrasts with phosphorus and arsenic oxyacids, which are invariably four coordinate), and all contain N–O and sometimes N–N $p\pi$–$p\pi$ bonding, which is favourable for the smaller p-block element. The anions derived from the main oxyacids are shown in Figure 7.12.

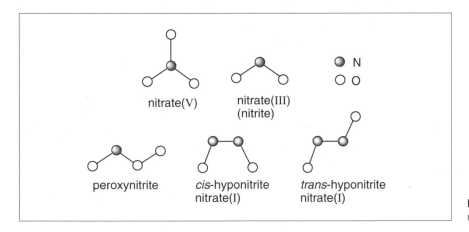

Figure 7.12 The shapes of the main nitrogen oxyanions

Nitric(V) acid is commercially the most important, and is manufactured on a large scale by the oxidation of NH_3 to NO under platinum-catalysed conditions (equation 7.11).

$$4NH_3 + 5O_2 \rightarrow 4NO + 6H_2O \qquad (7.11)$$

Nitrogen monoxide is then converted to nitric acid by air and water (equations 7.12 and 7.13).

$$2NO + O_2 \rightarrow 2NO_2 \qquad (7.12)$$

$$3NO_2 + H_2O \rightarrow 2HNO_3 + NO \qquad (7.13)$$

Lower oxidation state acids are typically unstable, though their conjugate anions are often stable. These include **nitric(III) acid** (nitrous acid,

HNO$_2$ is commonly used in organic chemistry, where it reacts with phenylamine derivatives RNH$_2$ to give diazonium salts RN$_2^+$, used in the manufacture of azo compounds RN=NR', many of which are dyes. HNO$_2$ solutions decompose on warming to give nitric acid, nitrogen monoxide and water.

HNO$_2$), formed by acidification of the nitrite anion (NO$_2^-$), and peroxonitric(III) acid (pernitrous acid, HOONO), formed by reaction of nitrous acid with hydrogen peroxide. A fourth acid, nitric(I) acid (hyponitrous acid, H$_2$N$_2$O$_2$), formed by reaction of silver nitrate(I) with dry HCl, readily decomposes to N$_2$O, but it can be stabilized in crystalline form by hydrogen bonding. The nitrate(I) anion itself is formed from reduction of NaNO$_2$ with Na/Hg, and has a *trans* configuration. In contrast, nitrate(I) formed by an alternative reaction of Na$_2$O with nitrous oxide, N$_2$O, has the *cis* structure (Figure 7.12).

7.6.2 Oxyacids of Phosphorus

A diverse range of phosphorus acids exist, with phosphorus in the +1, +3 and +5 oxidation states. The lower oxidation state compounds contain either P–P or (non-acidic) P–H bonds. Phosphoric(V) acid, H$_3$PO$_4$, commonly known as orthophosphoric acid, is the most well known, and is triprotic, all three hydrogens being able to be replaced by metals to give dihydrogenphosphate (H$_2$PO$_4^-$), monohydrogenphosphate (HPO$_4^{2-}$) and phosphate (PO$_4^{3-}$) salts. A diverse range of condensed phosphates exist, containing P–O–P linkages (see Section 12.2).

Phosphoric acid is manufactured industrially by dissolving P$_4$O$_{10}$ (Section 7.4.2) in boiling dilute phosphoric acid (see Worked Problem 2.3). An increasingly used, and less energy-intensive, route uses an acid displacement reaction with sulfuric acid (equation 7.14).

$$Ca_3(PO_4)_2 + 3H_2SO_4 + 6H_2O \rightarrow 2H_3PO_4 + 3CaSO_4.2H_2O \tag{7.14}$$

Phosphonic acid
(Phosphorous acid)
HP(O)(OH)$_2$

Dibasic

Phosphonic acid (H$_3$PO$_3$) contains one non-exchangeable hydrogen and is manufactured by the hydrolysis of PCl$_3$ (equation 7.15).

$$PCl_3 + 3H_2O \rightarrow H_3PO_3 + 3HCl \tag{7.15}$$

Phosphinic acid
(Hypophosphorous acid)
H$_2$P(O)OH

Monobasic

Phosphinic acid, H$_3$PO$_2$, contains two non-exchangeable hydrogens, and is made industrially by reaction of white phosphorus with sodium hydroxide to give the anion H$_2$PO$_2^-$, followed by acidification (equations 7.16 and 7.17). Both H$_3$PO$_3$ and H$_3$PO$_2$ are strong reducing agents, because they formally contain phosphorus in the +3 and +1 oxidation states, respectively.

$$P_4 + 4H_2O + 4OH^- \rightarrow 4H_2PO_2^- + 2H_2 \tag{7.16}$$

$$H_2PO_2^- + H^+ \rightarrow H_3PO_2 \tag{7.17}$$

Worked Problem 7.8

Q What is the oxidation state of phosphorus in the phosphorus acids (a) $H_4P_2O_5$ and (b) $H_2P_2O_6$?

A (a) $H_4P_2O_5$ is the anhydride formed by condensation of two OH groups from phosphonic acid molecules, H_3PO_3, giving a P–O–P linkage. The oxidation state is the same as in H_3PO_3, *i.e.* +3.
(b) $H_2P_2O_6$ has a P–P bond, which does not contribute to the oxidation state. The oxidation state is therefore one less than in phosphoric acid, H_3PO_4, *i.e.* +4.

$H_4P_2O_5$

$H_2P_2O_6$

7.6.3 Oxyacids of Arsenic, Antimony and Bismuth

Arsenic acid, H_3AsO_4, analogous to orthophosphoric acid, is made by dissolving elemental arsenic in nitric acid; it is also a triprotic acid, but is moderately oxidizing in *acid* solution. Arsenites (in the +3 oxidation state) are well known, such as in H_3AsO_3, but in contrast to the P–H bond in phosphites they contain no As–H bond.

For the heavier members of Group 15 there is little resemblance to the chemistry of phosphorus and arsenic; by comparison, antimonates and bismuthates are less well-defined compounds.

7.7   Halides

7.7.1 Pentavalent Compounds

The pentafluorides MF_5 are all known and stable, with the exception of NF_5 (which does not exist) since nitrogen only has four valence orbitals available. Fluoride ligands are able to stabilize the highest oxidation state of the Group 15 elements. The MF_5 halides are powerful fluoride acceptors (with the exception of BiF_5), giving the stable MF_6^- anions, and this property is widely used, for example in studies of noble gas chemistry (see Chapter 10). Antimony can form a dimeric ion $Sb_2F_{11}^-$, which has two SbF_6 octahedra joined by a common bridging fluoride (see Figure 10.6). The pentachlorides PCl_5 and $SbCl_5$ are, by contrast, well known, but $AsCl_5$ is a rather unstable substance. While PBr_5 is also known, other combinations, particularly involving the heavier Group 15 members and the heavier halides, are unstable owing to the combination of a Group 15 element in a high oxidation state and a reducing halide.

In the gas phase the pentahalides adopt trigonal bipyramidal geometries, as expected from VSEPR (see Section 1.4). In the solid state, however, the situation is more complicated, as illustrated by several phases of PCl_5 being characterized, none of which contain molecular PCl_5. PCl_5 crystallizes as an ionic solid $PCl_4^+PCl_6^-$, though a metastable form has been found to have the composition $(PCl_4^+)_2(PCl_6^-)Cl^-$. In contrast, solid PBr_5 contains PBr_4^+ and Br^- ions. A number of mixed halide species are also known, including the full range of EF_nCl_{5-n} species of P and As, *e.g.* PF_3Cl_2. The compounds EF_3Cl_2 exist as trigonal-bipyramidal molecular substances in the gas phase, but in the solid form they ionize to $ECl_4^+EF_6^-$, so that the less electron-withdrawing halogen is contained within the cationic species to minimize destabilization.

Worked Problem 7.9

Q PCl_5 crystallizes as $PCl_4^+PCl_6^-$. Predict the shapes of these ions.

A Phosphorus is in Group 15, with 5 valence electrons:

	PCl_4^+	PCl_6^-
central P atom	5 electrons	5 electrons
bonds to Cl atoms	4 electrons	6 electrons
charge	subtract 1 electron	add 1 electron
total	8 valence electrons	12 valence electrons
	4 pairs	6 pairs
shape	regular tetrahedron	regular octahedron

PCl_4^+

PCl_6^-

● P

○ Cl

7.7.2 Trivalent Compounds

All combinations of trivalent Group 15 elements E with halides X (EX_3) are known. This contrasts with the pentavalent species EX_5, which become increasingly oxidizing on going down the group. There are the usual gradations in properties on descending the group for a given halide. NF_3, PF_3 and AsF_3 exist as molecular substances, whereas SbF_3 is polymeric with fluoride bridges (reflecting the tendency for the heavier elements to have higher coordination numbers) and BiF_3 is ionic. NF_3 is the only stable nitrogen(III) halide; NCl_3 and NI_3 are highly explosive.[5]

PF_3 is one of the more widely studied trihalides of the Group 15 elements, because of its similarity to carbon monoxide (CO) in complexes with transition metals (see Section 6.7.1). Both ligands are able to act as σ-donors and as π-acceptors, and a wide range of analogous compounds such as $Ni(PF_3)_4$ and $Ni(CO)_4$ are known.

PF_3 is a poor σ-donor because of the pronounced electron-withdrawing effect of the three fluorine atoms; it is, however, a powerful π-acceptor ligand.

7.7.3 Oxyhalides

While trivalent oxyhalides of nitrogen are known, *e.g.* NOCl (Section 7.4.1), related compounds of phosphorus and arsenic are restricted to the pentavalent state. Thus, phosphorus and arsenic form many oxyhalides of the type MOX_3, which have approximately tetrahedral shapes. $POCl_3$ is manufactured on a large scale industrially, either by reaction of P_4O_{10} with PCl_5 (equation 7.18) or, increasingly, by oxidation of PCl_3 (equation 7.19).

O, S or Se

P

halogen

$$P_4O_{10} + 6PCl_5 \rightarrow 10POCl_3 \qquad (7.18)$$

$$2PCl_3 + O_2 \rightarrow 2POCl_3 \qquad (7.19)$$

7.8 Nitrides and Phosphides

Certain reactive metals, such as lithium and the Group 2 metals, form nitrides which appear to contain the nitride N^{3-} ion, because on hydrolysis ammonia is produced, *e.g.* equation 7.20.

$$Li_3N + 3H_2O \rightarrow NH_3 + 3Li^+ + 3OH^- \qquad (7.20)$$

In contrast, compounds of other semi-metallic elements such as boron are viewed as covalently bonded compounds.

Phosphorus and arsenic similarly form phosphides and arsenides, which give PH_3 and AsH_3 on hydrolysis; this is one method of synthesizing PH_3, and others are described in Sections 7.3.1 and 2.4.2.

The N atom in N^{3-} has a complete octet of electrons, and is isoelectronic with F^- and O^{2-}, but is highly reactive due to the three negative charges.

Boron nitride (BN) is discussed in Section 5.8.2.

Summary of Key Points

1. *Properties:* going down the group, there is a gradation from non-metallic to metallic character similar to Group 14.

2. The *group oxidation state of +5 occurs for all elements*, and owing to the inert pair effect the heavier elements Sb and Bi form stable compounds in the +3 oxidation state.

3. *Nitrogen forms many molecular oxides*, stabilized by strong $p\pi$–$p\pi$ bonding.

Problems

7.1. What is the oxidation state of the Group 15 element in the following compounds or ions: (a) N_2H_4; (b) NO_3^-; (c) AsF_6^-; (d) H_3PO_2; (e) 'phosphorus pentoxide'.

7.2. Identify the element X in each of the following:
(a) X forms more well-characterized oxides than any other element, and exists only in one allotrope.
(b) X forms acids H_3XO_n where n is 2, 3 or 4.
(c) The heavier trivalent halides of X are explosive and the pentavalent halides do not exist.
(d) The hydride XH_3 is very unstable; X is not radioactive.

7.3. Balance the following chemical equations:
(a) $NH_4NO_{2(s)} \rightarrow N_{2(g)} + H_2O_{(g)}$
(b) $NH_4NO_{3(s)} + \text{heat} \rightarrow$
(c) $Zn_3As_{2(s)} + HCl_{(aq)} \rightarrow ZnCl_{2(aq)} + AsH_{3(g)}$
(d) $As_2O_3 + Zn \rightarrow AsH_3$ (in acid solution)
(e) $P_4O_{10} + \text{excess boiling } H_2O \rightarrow$

7.4. Plot a graph of boiling point versus molecular mass for the Group 15 hydrides, and comment on the shape of the curve. Boiling points (°C): NH_3 (–33), PH_3 (–88), AsH_3 (–55), SbH_3 (–17).

7.5. Explain why the X–N–O bond angles for the nitrosyl halides ONX vary in the series F (110°), Cl (116°) and Br (117°).

7.6. Single, double and triple bond strengths for P–P bonds are given in Table 7.1. Show that these are consistent with the occurrence of phosphorus in singly bonded forms.

References

1. K. O. Christe, W. W. Wilson, J. A. Sheehy and J. A. Boatz, *Angew. Chem., Int. Ed. Engl.*, 1999, **38**, 2004.
2. H. Hartl, *Angew. Chem., Int. Ed. Engl.*, 1995, **34**, 2637.
3. A. R. Butler, *Chem. Br.*, 1990, 419; A. R. Butler and D. L. H. Williams, *Chem. Soc. Rev.*, 1993, **22**, 233; R. J. P. Williams, *Chem. Soc. Rev.*, 1996, **25**, 77.
4. J. Clade, F. Frick and M. Jansen, *Adv. Inorg. Chem.*, 1994, **41**, 327.
5. I. Tornieporth-Oetting and T. Klapötke, *Angew. Chem., Int. Ed. Engl.*, 1990, **29**, 677.

Further Reading

The structures of the group 15 element(III) halides and halogenoanions, G. A. Fisher and N. C. Norman, *Adv. Inorg. Chem.*, 1994, **41**, 233.

Recent developments in the chemistry of covalent azides, T. M. Klapötke, *Chem. Ber./Recueil*, 1997, **130**, 443.

The chemistry of diphosphenes and their heavy congeners: synthesis, structure and reactivity, L. Weber, *Chem. Rev.*, 1992, **92**, 1839.

The nitrogen fluorides and some related compounds, H. J. Eméleus, J. M. Shreeve and R. D. Verma, *Adv. Inorg. Chem.*, 1989, **33**, 139.

Covalent inorganic azides, I. C. Tornieporth-Oetting and T. M. Klapötke, *Angew. Chem., Int. Ed. Engl.*, 1995, **34**, 511.

Nitric oxide - some old and new perspectives, E. W. Ainscough and A. M. Brodie, *J. Chem. Educ.*, 1995, 686.

8

The Group 16 (Chalcogen) Elements: Oxygen, Sulfur, Selenium, Tellurium and Polonium

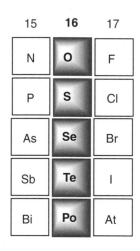

Catenation of carbon: see Section 6.1.

Aims

By the end of this chapter you should understand:

- The marked differences between the chemistry of oxygen and sulfur, and the diversity of the chemistry in this group
- The strong tendency of sulfur to form polysulfur chains, in the element and in some compounds

8.1   Introduction and Oxidation State Survey

The Group 16 elements are the last group to contain a true metal, polonium. However, the same general periodic trends are maintained as in the previous two groups, with the characteristics of the elements ranging from true non-metals (oxygen and sulfur), through semiconductors (selenium and tellurium) to metallic polonium. The formation of rings and chains of an element, 'catenation', is an important aspect of sulfur chemistry in particular, and also occurs in some compounds of selenium and tellurium, but hardly ever in oxygen compounds. The group oxidation state is +6, and while this is the stable oxidation state for sulfur, it does not exist for oxygen (which is restricted to a maximum positive oxidation state of +2 in OF_2) and becomes increasingly oxidizing on going from Se to Te to Po. There is the usual (for p-block elements) stabilization of lower oxidation states going down the group.

> ### Worked Problem 8.1
>
> **Q** In many cases, sulfur analogues of oxygen compounds are known, often indicated in the name by the **thio** prefix. Name the following pairs of related oxygen and sulfur species: (a) SO_4^{2-} and $S_2O_3^{2-}$; (b) OCN^- and SCN^-; (c) $(H_2N)_2C=O$ and $(H_2N)_2C=S$.
>
> **A** (a) sulfate and thiosulfate; (b) cyanate and thiocyanate; (c) urea and thiourea.

8.2   The Elements

8.2.1 Oxygen

Oxygen is the second most abundant gas in the Earth's atmosphere (see Figure 7.2) and it is obtained industrially from liquid air by fractional distillation. The most important **allotrope** of oxygen is O_2, which is blue in the liquid and solid states. Oxygen is used directly in combustion processes, or reacted with alkenes to give epoxides which are used in the manufacture of surfactants. Its compounds often find uses as sterilizing agents and bleaches.

The bonding in the O_2 molecule is discussed in Section 1.5.

Allotropes and polymorphs: Section 6.2.1.

Exposure of oxygen to an electric discharge converts it into **ozone**, O_3, which is a diamagnetic, but extremely reactive, non-linear allotrope. Ozone is an extremely important molecule in the stratosphere, where it is produced by the reaction of oxygen atoms (formed by dissociation of O_2 molecules, equation 8.1) with O_2 (equation 8.2).

Ozone, O_3, is isoelectronic with the nitrite anion, NO_2^-, since N^- has the same number of electrons as an O atom.

$$O_2 + h\nu \rightarrow 2O\bullet \tag{8.1}$$

$$O_2 + O\bullet \rightarrow O_3 \tag{8.2}$$

$$O_3 + h\nu \rightarrow O_2 + O\bullet \tag{8.3}$$

The ozone itself is photolysed (dissociated) by UV radiation (equation 8.3), at wavelengths which O_2 cannot absorb, thus giving protection from the harmful effects of short-wavelength UV.

8.2.2 Sulfur

Sulfur occurs naturally in several forms: as the element, in sulfate minerals such as *gypsum* ($CaSO_4.2H_2O$) and in sulfide minerals (for example *pyrite*, FeS_2). Elemental sulfur occurs in underground deposits, and

In the Frasch process, superheated water is passed down a borehole, which melts the sulfur and allows it to be pumped to the surface by compressed air.

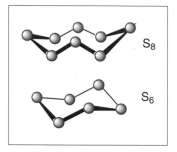

Figure 8.1 Some sulfur rings; S_8 is much more common under normal conditions

S_2 is analogous to dioxygen, O_2, and can be condensed to form a purple paramagnetic solid.

is extracted by the **Frasch process**. It is also obtained in relatively large amounts from the processing of crude oil.

Sulfur exists in numerous *allotropes*, with chain and ring structures; indeed, sulfur probably forms more allotropes and polymorphs than any other element. The most common form is **rhombic sulfur**, with S_8 rings (Figure 8.1). There is also a **monoclinic polymorph** of S_8 rings, which has a different packing arrangement in the crystal. Molten sulfur contains a myriad of ring and chain forms of sulfur, of different sizes, depending on the conditions. At the melting point (*ca.* 115 °C), sulfur is a yellow, low-viscosity liquid comprising mainly S_8 rings, with traces of other rings ranging in size from S_6 (Figure 8.1) to S_{30} or more. Further heating initially decreases the viscosity of the molten sulfur, but then it increases above *ca.* 159 °C, reaching a maximum at 170 °C, as high molecular mass rings and chains are formed. At the boiling point (444 °C), sulfur vapour consists mainly of S_7 (40%), S_6 (30%) and S_8 (20%), with smaller amounts of S_4, S_5 and S_2.

Worked Problem 8.2

Q The enthalpy change for the conversion of $S_{8(g)}$ into $4S_{2(g)}$ is +401 kJ, and the average bond energy in the S_8 molecule is +210.8 kJ mol^{-1}. Calculate the bond energy (E) of the S_2 molecule. Is this value consistent with S_8 being the stable room temperature allotrope of sulfur?

A Set up a thermodynamic cycle:

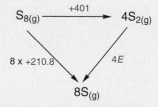

Thus, $4E + 401 = (8 \times 210.8)$; hence $E = +321$ kJ mol^{-1}. This value is less than twice that of the S–S bond in S_8, so one S=S double bond is less stable with respect to two S–S single bonds, and therefore S_8 is the more stable allotrope.

8.2.3 Selenium, Tellurium and Polonium

Grey (crystalline) selenium and tellurium contain long spiral molecules with weak metallic interactions between adjacent chains, which break down into smaller rings and chains on melting.

Polonium occurs naturally as a trace component of uranium ores, but is made artificially by the irradiation of ^{209}Bi with neutrons (equation 8.4).

$$^{209}\text{Bi} + \text{n} \rightarrow {}^{210}\text{Bi} \rightarrow {}^{210}\text{Po} + \beta \qquad (8.4)$$

8.3 Hydrides and Related Species

8.3.1 Compounds of the Type H$_2$E

These are known for all the Group 16 elements, but become increasingly unstable on going down the group, as shown by their enthalpies of formation (Table 8.1). This instability is paralleled by compounds such as the EH$_3$ series (E = N to Bi; see Section 7.3.1) and the HE′ series (E′ = F to I; see Section 9.4).

Nomenclature: the IUPAC names for H_2O and H_2S are oxane and sulfane, respectively.

Laboratory synthesis:
$M_2E + 2HCl \rightarrow H_2E + 2MCl$
(e.g. $M_2E = Na_2S, Na_2Se$).

Table 8.1 Enthalpies of formation of the hydrides H$_2$E.

H_2E	$\Delta_f H°$ (kJ mol^{-1})
H$_2$O	−241.8
H$_2$S	−20.1
H$_2$Se	+85.8
H$_2$Te	+154.4

The boiling points of the H$_2$E compounds reveal the anomalously high boiling point for water, owing to very strong hydrogen bonding (see Figure 2.4 and Section 2.6.2). Water is, of course, an extremely familiar substance, with complex structures involving extensive hydrogen bonding in both the solid and solution states. In marked contrast to water, H$_2$S, H$_2$Se and H$_2$Te are highly toxic, foul-smelling gases.

Water undergoes self-ionization (equation 8.5), with an equilibrium constant of 10^{-14} at 25 °C. Acids are substances which increase the concentration of H$_3$O$^+$ (Section 8.3.3) while bases increase the concentration of OH$^-$.

$$2H_2O \rightleftharpoons H_3O^+ + OH^- \qquad (8.5)$$

Worked Problem 8.3

Q How would you expect the boiling points of the organic derivatives Me_2E (E = O, S, Se or Te) to compare with those of the corresponding hydrides, H_2E?

A In the methyl derivatives, specifically in Me_2O, no hydrogen bonding is possible, because hydrogen is not bonded to the oxygen. Thus, a relatively smooth increase in boiling point with molecular mass is expected. This contrasts with the boiling points of the hydrides (see Figure 2.4), where water has an anomalously high boiling point due to hydrogen bonding.

8.3.2 Chalcogenide Anions

All the chalcogens form chalcogenide anions E^{2-} (*i.e.* oxides, sulfides, selenides and tellurides). These are formed by double deprotonation of H_2E; a corresponding series of anions EH^- also exist, typified by the well-known sodium hydroxide, NaOH. Experimentally, chalcogenide salts are usually formed by direct combination of the elements (sometimes using a non-aqueous solvent such as liquid ammonia), *e.g.* equation 8.6.

$$2Na + Se \rightarrow Na_2Se \qquad (8.6)$$

By forming these ions the chalcogen attains a noble gas electronic configuration, and the E^{2-} ions are isoelectronic with the corresponding halide anions (also with the nitride N^{3-} and phosphide P^{3-} anions). The chalcogenide anions become increasingly unstable going down the group, because the increasing size means the electrons are less tightly held, so oxidation occurs more readily. Many compounds containing E^{2-} ions are water sensitive, especially for Group 1 and 2 metals, and the degree of hydrolysis is greater for Se and Te (equations 8.7 and 8.8).

$$Na_2S + H_2O \rightarrow NaSH + NaOH \qquad (8.7)$$

$$Na_2Te + 2H_2O \rightarrow H_2Te + 2NaOH \qquad (8.8)$$

Sulfides and selenides of soft, polarizable, heavy metals (*e.g.* Cd, Pb, Hg) often occur naturally, are moisture stable, and precipitate readily when H_2S or H_2Se is passed through an aqueous metal ion solution (equation 8.9).

$$Pb^{2+}_{(aq)} + H_2S_{(g)} \rightarrow PbS_{(s)} + 2H^+_{(aq)} \tag{8.9}$$

There is a vast solid-state chemistry of metal oxides, sulfides and selenides.

8.3.3 Cationic Onium (H₃E⁺) Ions

The oxonium ion, H_3O^+, is well known; see Section 2.5. The sulfonium and selenonium cations H_3S^+ and H_3Se^+ can also be made (despite the lower basicity of H_2S and H_2Se compared to water, which makes protonation more difficult) by using the very strong acid $H^+SbF_6^-$ (generated from HF and SbF_5) (equation 8.10).

$$H_2S + HF + SbF_5 \rightarrow H_3S^+SbF_6^- \tag{8.10}$$

Organic derivatives of these sulfonium salts are widely known; Me_3O^+ salts are reactive, and widely used as reagents for transferring a Me^+ group.

8.3.4 Hydrogen Peroxide and Peroxides

Hydrogen peroxide, H_2O_2, is the most well-known peroxide. The molecule adopts a *gauche* structure both in the gas phase and in the solid state (Figure 8.2), owing to repulsion of the lone pairs on the oxygen atoms. Hydrogen peroxide undergoes a similar self-ionization to water (equation 8.5), but to a slightly greater extent (equation 8.11).

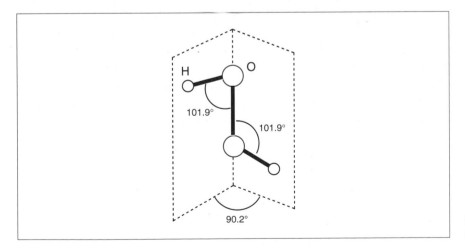

Figure 8.2 The structure of hydrogen peroxide, H_2O_2

$$2H_2O_2 \rightleftharpoons H_3O_2^+ + HO_2^- \tag{8.11}$$

The equilibrium constant is 1.5×10^{-12}, whereas for water it is 10^{-14}.

The $H_3O_2^+$ ion is formed by protonation of H_2O_2, just as H_3O^+ is formed by protonation of water.

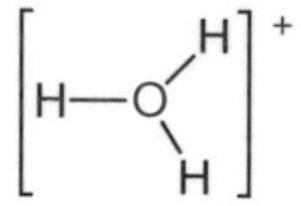

$$\left[\begin{array}{c} H \\ H-O \\ H \end{array} \right]^{+}$$

The H_3O^+ ion

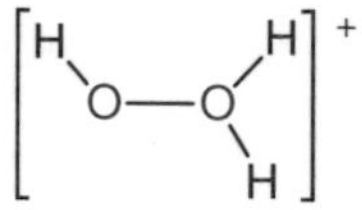

$$\left[\begin{array}{c} H \\ O-O \\ H \end{array} \right]^{+}$$

The $H_3O_2^+$ ion

Hydrogen peroxide is *stable* towards decomposition into *hydrogen and oxygen*, as shown by the large, negative enthalpy of formation ($\Delta_f H°$ -187.8 kJ mol^{-1}); however, it is *unstable* towards decomposition to *water and oxygen* (equation 8.12).

$$2H_2O_2 \rightarrow 2H_2O + O_2 \qquad \Delta H = -98.3 \text{ kJ mol}^{-1} \qquad (8.12)$$

Box 8.1 Synthesis and Uses of Hydrogen Peroxide

Most hydrogen peroxide is manufactured industrially by the anthraquinone autoxidation process, where an anthraquinol dissolved in an organic solvent is oxidized (using air) to the corresponding anthraquinone, plus hydrogen peroxide (Figure 8.3). The hydrogen peroxide is recovered by extracting the organic solution with water. The anthraquinone is then reduced back to the anthraquinol using hydrogen and a palladium catalyst. This process is repeated many times, giving a catalytic cycle.

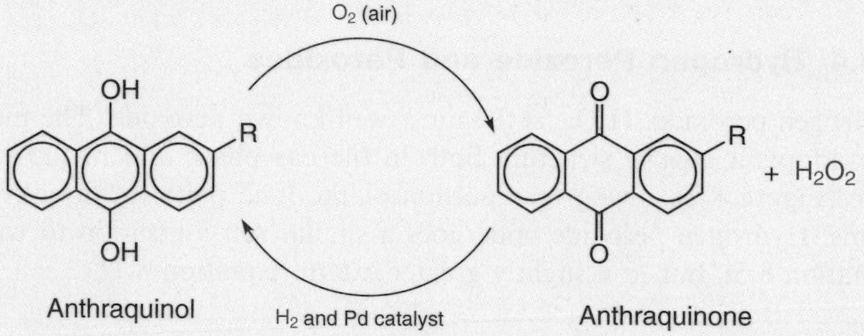

Hydrogen peroxide finds various uses, two of which are as bleaching and sterilizing agents.

Figure 8.3 The anthraquinol–anthraquinone process for hydrogen peroxide manufacture

In the laboratory, hydrogen peroxide can be synthesized by the reaction of barium peroxide (BaO_2) with sulfuric acid, and then removing the insoluble barium sulfate precipitate (equation 8.13).

$$BaO_{2(s)} + H_2SO_{4(aq)} \rightarrow H_2O_{2(aq)} + BaSO_{4(s)} \qquad (8.13)$$

Alkali metal peroxides: Section 3.5.

Peroxide salts, containing the singly bonded O_2^{2-} anion, are formed by the Group 1 and Group 2 metals, and examples are barium peroxide (BaO_2; Box 8.1) and sodium peroxide (Na_2O_2). Hydroperoxides (containing the OOH$^-$ anion) are also formed by the Group 1 metals, such as NaOOH.

All compounds which contain peroxo groups are oxidizing agents, and this is the main feature of their chemistry. No compounds are known which have greater than two oxygens in a chain terminated by hydrogens (such as HOOOH), but related fluorine-containing compounds such as FOOOOF and CF_3OOOCF_3 are known.[1]

Worked Problem 8.4

Q Which of the following compounds contain the peroxide ion: (a) barium peroxide, BaO_2; (b) hydrogen peroxide, H_2O_2; (c) potassium peroxide, K_2O_2; (d) lead(IV) oxide, PbO_2.

A BaO_2 and K_2O both contain the O_2^{2-} ion. Hydrogen peroxide, H_2O_2, is a covalent substance, while PbO_2 is not a peroxide but is the dioxide of lead(IV).

8.3.5 Sulfur, Selenium and Tellurium Hydrides and Anions with Chalcogen–Chalcogen Bonds

A number of hydrides are known which contain one or more –S–S–, –Se–Se– or –Te–Te– linkages. The longest chains are found for sulfur, with compounds of composition H_2S_n being stable for up to 8 sulfur atoms, and possibly more. This is not unexpected, since elemental sulfur itself has a strong tendency to form long chains (catenation); see Section 8.2.2. A mixture of polysulfanes can be prepared by the addition of acid to a polysulfide solution (equation 8.14).

$$(S_n)^{2-} + 2H^+ \rightarrow H-(S)_n-H \quad \text{(polysulfane mixture, } n = 2\text{–}6)$$
(8.14)

Compounds of the type H_2E_x (E = S, Se or Te) tend to be unstable, giving the element and H_2E (equation 8.15).

$$H-(E)_n-H \rightarrow H_2E + (n-1)E$$
(8.15)

Longer-chain polychalcogenide anions of S, Se and Te are generally prepared by reaction of an alkali metal (M) with the element (using a non-aqueous solvent, *e.g.* liquid ammonia) or from the salt (M_2E) by reaction with the chalcogen (E), *e.g.* equations 8.16 and 8.17.

$$4Na + 5Te \rightarrow Na_2Te_2 + Na_2Te_3$$
(8.16)

$$Cs_2S + 4S \rightarrow Cs_2S_5$$
(8.17)

8.4   Halides

8.4.1 Oxygen Halides

The six known oxygen fluorides:
OF_2
O_2F_2
O_3F_2
O_4F_2
O_5F_2
O_6F_2

Oxygen forms six fluorides, which have chains of from one to six oxygen atoms, terminated by fluorines. Since fluorine always has a negative oxidation state (being the most electronegative element), these oxygen fluorides are rare examples of compounds with oxygen in positive oxidation states.

OF$_2$ is the most stable compound, formed by reaction of fluorine with dilute sodium hydroxide solution (see Section 9.5.1). The other compounds are prepared by an electrical discharge on mixtures of O_2 and F_2 at low temperatures (*ca.* −185 °C). Reaction of O_2F_2 with a fluoride ion acceptor, such as BF_3 or PF_5, gives salts containing the dioxygenyl cation (O_2^+) (equation 8.18). The dioxygenyl cation is also obtained by direct oxidation of O_2 using platinum hexafluoride (PtF_6) (equation 8.19), which led to the discovery of the very first noble gas compound, by Bartlett (see Section 10.3.2).

$$2O_2F_2 + 2PF_5 \rightarrow 2(O_2)^+(PF_6)^- + F_2 \qquad (8.18)$$

$$O_2 + PtF_6 \rightarrow (O_2)^+(PtF_6)^- \qquad (8.19)$$

There is a wide range of other chlorine, bromine and iodine oxides known, for example Cl_2O, Cl_2O_7, I_2O_5, *etc.*, described in Chapter 9.

8.4.2 Sulfur Halides

Known sulfur halides

Oxidation state	Examples
<+1	S_nCl_2 (n = 3–8), S_nBr_2 (n = 3–8)
+1	S_2F_2, S_2Cl_2, S_2Br_2, S_2I_2
+2	SF_2, SCl_2
+4	SF_4, SCl_4
+5	S_2F_{10}
+6	SF_6

Sulfur forms a wide range of halides, particularly with fluorine, in oxidation states ranging from low, fractional positive oxidation states [in halosulfanes of the type $X–(S)_n–X$] through to the group oxidation state of +6 (in SF_6).

The so-called 'monohalides', S_2X_2, have structures similar to hydrogen peroxide (Figure 8.2), but S_2F_2 has been found to exist in two forms, one being the hydrogen peroxide type and the other having sulfur in two different oxidation states (Figure 8.4).

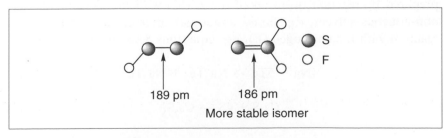

189 pm   186 pm

More stable isomer

● S
○ F

Figure 8.4 The two isomers of S_2F_2

The compounds S_2Cl_2 and S_2Br_2 are the smallest members of a series of chloro- and bromosulfanes of the type S_nX_2, which have sulfur–sulfur bonds and up to eight sulfur atoms. They are analogous to the polysulfanes $H(S)_nH$ themselves (Section 8.3.5), but with chain-terminating halogens instead of hydrogens. These compounds, like the sulfanes and the element itself, show the strong tendency of sulfur to form catenated species. S_2Cl_2 and S_2Br_2 are formed by direct combination of the elements (equation 8.20).

$$S_8 + 4Cl_2 \rightarrow 4S_2Cl_2 \qquad (8.20)$$

The only stable sulfur dihalide is the bent SCl_2, which disproportionates to give S_2Cl_2 and Cl_2 (equation 8.21). SF_2 is even less stable.

$$2SCl_2 \rightarrow S_2Cl_2 + Cl_2 \qquad (8.21)$$

Sulfur(IV) halides are restricted to SF_4 and SCl_4 (the latter being the less stable, and only occurring in the solid state where it is thought to exist in an ionic form as $SCl_3{}^+Cl^-$). SF_4 has a structure derived from a trigonal bipyramid; see Worked Problem 1.6. Although highly moisture-sensitive (equation 8.22), it is finding use as a fluorinating agent in organic chemistry.

Structure of SF_4:

$$SF_4 + H_2O \rightarrow SOF_2 + 2HF \qquad (8.22)$$

In contrast to the high reactivity of SF_2 and SF_4, sulfur hexafluoride (SF_6) is unreactive. The hydrolysis of SF_6 is *thermodynamically* very favourable, indicated by the large, negative $\Delta G°$ value (-301 kJ mol^{-1}) calculated for the hydrolysis reaction in equation 8.23.

Structure of SF_6:

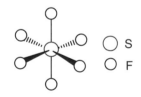

$$SF_{6(g)} + 3H_2O_{(g)} \rightarrow SO_{3(g)} + 6HF_{(g)} \qquad (8.23)$$

However, the hydrolysis (and other reactions of SF_6) does not proceed because of *kinetic* reasons, because the sulfur atom is sterically protected by the six fluorine atoms, preventing attack by an incoming nucleophile.

8.4.3 Selenium and Tellurium Halides

The halides of selenium and tellurium are similar to those of sulfur, but there are some notable differences:

- The tendency to form catenated compounds is markedly less for Se and Te.
- Selenium and particularly tellurium halides, especially in the +4 oxi-

Known selenium and tellurium halides:

Oxidation state	Examples
+1	Se_2Cl_2, Se_2Br_2
+2	$SeCl_2$, $SeBr_2$, $TeCl_2$, $TeBr_2$, TeI_2
+4	SeF_4, $(SeX_4)_4$ (X = Cl, Br), $(TeF_4)_x$, $(TeX_4)_4$ (X = Cl, Br, I)
+5	Te_2F_{10}
+6	SeF_6, TeF_6

dation state, may be polymeric. TeF_4 is polymeric, but for $SeCl_4$, $SeBr_4$, $TeCl_4$, $TeBr_4$ and TeI_4, tetramers $(MX_4)_4$ occur (Figure 8.5).

- Selenium and tellurium halides have a greater tendency to form complex anions by addition of halide anions. Thus, for example, $SeCl_4$ adds one or two chloride ions to form $SeCl_5^-$ or $SeCl_6^{2-}$ ions, respectively.

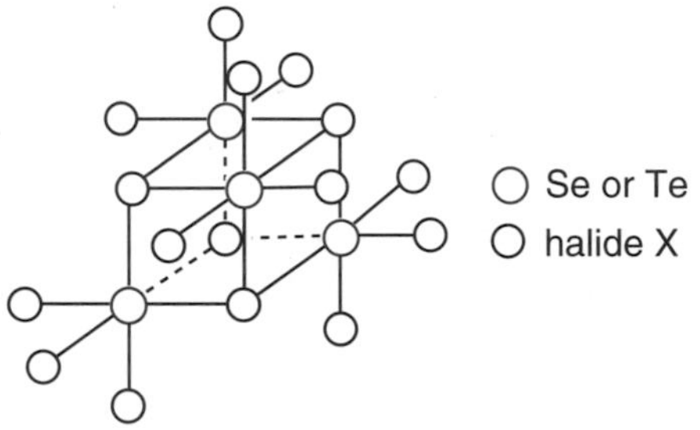

Legend: ○ Se or Te; ○ halide X

Figure 8.5 Structure of the tetrameric unit in the compounds $(MX_4)_4$ (M = Se, Te; X = halide)

8.5   Oxides

8.5.1 Dioxides, EO_2

The dioxides EO_2 are known for S, Se, Te and Po, and are formed by heating the elements in air, for example sulfur (equation 8.24).

$$S_8 + 8O_2 \rightarrow 8SO_2 \tag{8.24}$$

The dioxides have very different structures. SO_2 is a covalent molecular substance, SeO_2 has an infinite covalent chain structure in the solid state (and a molecular SO_2-type structure in the gas phase) (Figure 8.6), while TeO_2 and PoO_2 have ionic structures. The different structures reflect the increasing metallic character of the elements going down the group, and the tendency for only the light elements (Se and particularly S) to engage in E=O π-bonding. For SeO_2, some π-bonding is sacrificed to provide extra σ-bonding; the same reasons account for the different structures of (molecular) CO_2 and (polymeric) SiO_2 in Group 14.

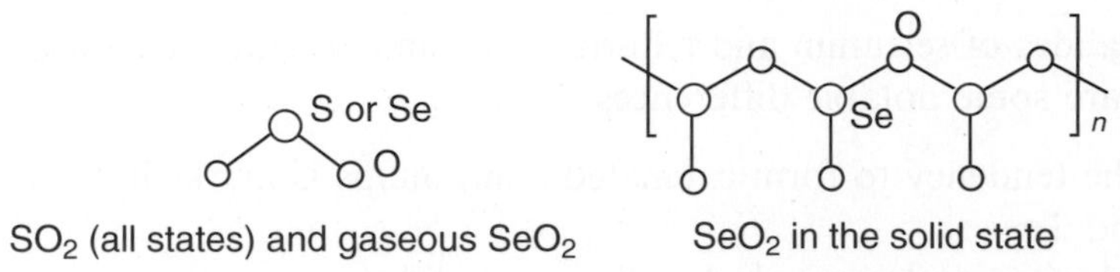

SO_2 (all states) and gaseous SeO_2 SeO_2 in the solid state

Figure 8.6 Structures of SO_2 and SeO_2

8.5.2 Trioxides, EO_3

The trioxides are known for S, Se and Te. SO_3 in the gas phase is a trigonal-planar molecule. In the solid state, SO_3 forms several different structures (Figure 8.7), where some S=O π-bonding has been sacrificed for additional σ-bonding. SeO_3 is more strongly oxidizing than SO_3.

SO_3 is manufactured on a huge scale industrially by the V_2O_5-catalysed oxidation of SO_2; most of the SO_3 is converted into sulfuric acid (see Section 8.7.2).

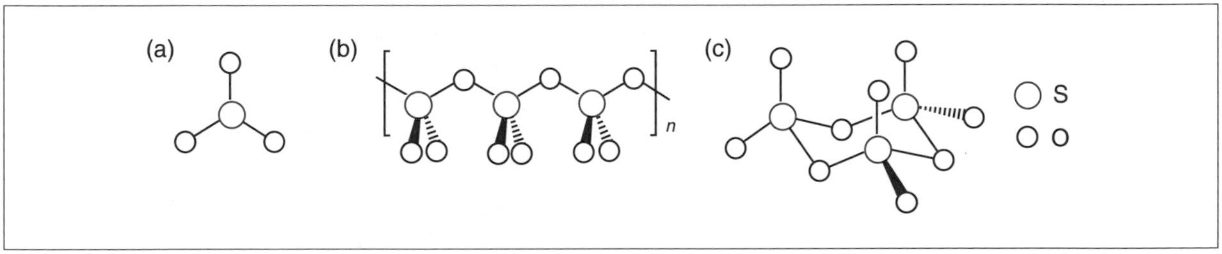

8.6   Oxyhalides of Sulfur, Selenium and Tellurium

Figure 8.7 Structures of SO_3: (a) in the gas phase; (b, c) in the solid state

The most important compounds are the hexavalent compounds of the type EO_2X_2 and the tetravalent compounds EOX_2. The sulfuryl halides (or more systematically, sulfur dihalide dioxides) SO_2X_2 (known for X = chlorine or fluorine) and the selenyl fluoride SeO_2F_2 are known. Thionyl halides (sulfur dihalide oxides) of the type SOX_2 are known for fluorine, chlorine and bromine, and the selenium analogues $SeOX_2$ also exist. All sulfuryl and thionyl halides, and their selenium analogues, are powerful halogenating agents and find applications in inorganic and organic chemistry. Examples include the preparation of chlorides from alcohols (equation 8.25) and the preparation of anhydrous metal halides (equation 8.26).

$$R–OH + SOCl_2 \rightarrow R–Cl + SO_2 + HCl \qquad (8.25)$$

$$NiCl_2.6H_2O_{(s)} + \text{excess } SOCl_{2(l)} \rightarrow NiCl_{2(s)} + 6SO_{2(g)} + 12HCl_{(g)} \qquad (8.26)$$

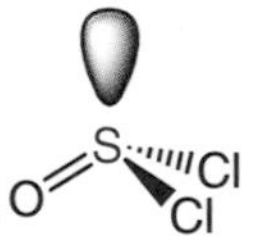

Sulfur dichloride oxide (Thionyl chloride) $SOCl_2$
Trigonal pyramidal

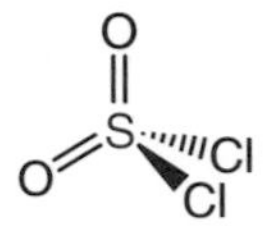

Sulfur dichloride dioxide (Sulfuryl chloride) SO_2Cl_2
Tetrahedral

Worked Problem 8.5

Q Predict the shapes of the molecules $SOCl_2$ and SO_2Cl_2.

A Applying VSEPR (see Section 1.4), S has 6 valence electrons. In $SOCl_2$, the S has 6 + 3 electrons (σ-bonds to Cl and O), −1 electron (for S=O π-bond), which gives 8 electrons or 4 pairs. Thus $SOCl_2$ has a trigonal pyramidal shape. In SO_2Cl_2 the S also has 4 pairs of electrons, and since there are 4 atoms the molecule is tetrahedral. In both $SOCl_2$ and SO_2Cl_2 the Cl–S–Cl bond angle will be decreased from the regular tetrahedral bond angle of 109.5° because of the effect of the lone pair and double bond(s).

8.7 Oxyacids of Sulfur, Selenium and Tellurium

A very wide range of oxyacids and/or their anions is known for the Group 16 elements, and the widest range is formed by sulfur. Sulfur and selenium can have a coordination number of up to four, while tellurium tends to be six coordinate. Compounds with S–S bonds are also known.

8.7.1 Sulfurous, Selenous and Tellurous acids, H_2EO_3

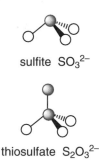

sulfite SO_3^{2-}

thiosulfate $S_2O_3^{2-}$

● S
○ O

tetrathionate $S_4O_6^{2-}$

Sulfur dioxide is very soluble in water and dissolves largely without reaction, forming an acidic solution commonly referred to as 'sulfurous acid' (H_2SO_3). The free acid has never been isolated, and is either absent or present in only trace amounts in aqueous SO_2 solutions.[2]

Selenium dioxide and tellurium dioxide behave similarly, although the solutions are less acidic than H_2SO_3. A solution of SO_2, SeO_2 or TeO_2 in aqueous hydroxide solutions gives EO_3^{2-} or HEO_3^- anions. Sulfite reacts with sulfur on heating to give the well-known thiosulfate anion (equation 8.27).

$$SO_3^{2-} + S \rightarrow S_2O_3^{2-} \qquad (8.27)$$

Thiosulfate is widely used in redox titrations involving iodine (see equation 9.28). In this reaction the tetrathionate ion ($S_4O_6^{2-}$) is formed, which contains a chain of four S atoms. Tetrathionate is one member of a series of (unstable) sulfur-bridged polythionates of composition $^-O_3S{-}S_n{-}SO_3^-$ where n can be any value from 0 to 22.

8.7.2 Sulfuric, Selenic and Telluric Acids

The purpose of dissolving the SO_3 in H_2SO_4 is to moderate the violent reaction which would occur between SO_3 and water.

Sulfuric acid is well known and is prepared by the oxidation of sulfur dioxide to sulfur trioxide (Section 8.5.2), followed by hydrolysis of the SO_3 in sulfuric acid to give pyrosulfuric acid (often called oleum), which is then hydrolysed (equations 8.28–8.30).

$$2SO_2 + O_2 \rightarrow 2SO_3 \qquad (8.28)$$

$$SO_3 + H_2SO_4 \rightarrow H_2S_2O_7 \text{ (pyrosulfuric acid)} \qquad (8.29)$$

$$H_2S_2O_7 + H_2O \rightarrow 2H_2SO_4 \qquad (8.30)$$

Sulfuric acid and the sulfate anion (SO_4^{2-}) are difficult to reduce, and sulfuric acid is a very strong acid ($pK_a < 0$).

Selenic acid, H_2SeO_4, is similar to sulfuric acid; however, the corresponding acid containing tellurium(VI) is the diprotic hexahydroxy acid, $Te(OH)_6$. This occurs formally by addition of water to Te=O groups, and thus tellurium is showing the characteristic increase in its coordination number observed for other heavy p-block elements, such as iodine (Section 9.6.4).

8.7.3 Other Sulfur Acids and their Anions

Many sulfur acids can formally be derived by substitution reactions of other sulfur acids, as shown in Figure 8.8. Thus, starting with sulfuric acid, one of the OH groups may be replaced by a hydroperoxo group (OOH) to give monoperoxosulfuric acid (Section 8.3.4). An OH group can also be replaced with a halogen, an NH_2 (or organic substituted RNH) group, or an organic group, *e.g.* Me. The thiosulfate anion $S_2O_3^{2-}$ has the same structure as sulfate (SO_4^{2-}) but with an oxygen replaced by a sulfur.

Other sulfur oxyanions, shown in Figure 8.9, are the dithionite ion ($S_2O_4^{2-}$) and the pyrosulfite anion ($S_2O_5^{2-}$); the next member of this series is the dithionate anion, $S_2O_6^{2-}$.

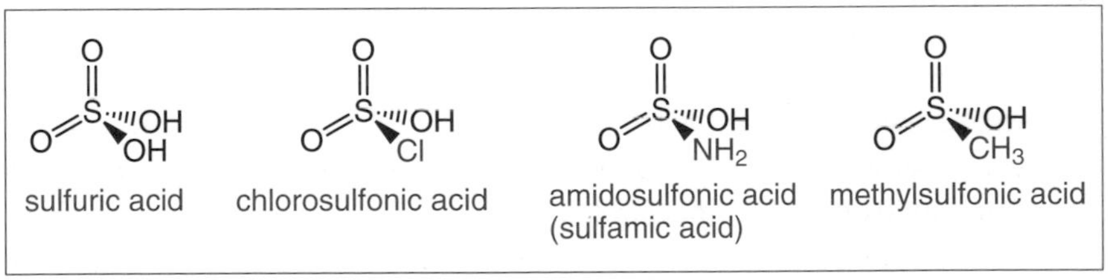

![Structures of sulfuric acid derivatives: sulfuric acid, chlorosulfonic acid, amidosulfonic acid (sulfamic acid), methylsulfonic acid]

sulfuric acid   chlorosulfonic acid   amidosulfonic acid (sulfamic acid)   methylsulfonic acid

Figure 8.8 Some sulfuric acid derivatives

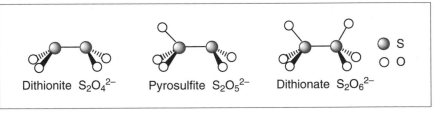

Dithionite $S_2O_4^{2-}$ Pyrosulfite $S_2O_5^{2-}$ Dithionate $S_2O_6^{2-}$

● S
○ O

Figure 8.9 Schematic relationship between some sulfur oxyanions

Worked Problem 8.6

Q What is the oxidation state(s) of sulfur in the following ions: (a) dithionite; (b) peroxodisulfate; (c) thiosulfate.

A (a) In the dithionite $S_2O_4^{2-}$ ion there is an S–S bond which does not contribute to the oxidation state. Hence the oxidation state of S is +3.
(b) Disregarding the structure of the ion, an oxidation state calculation gives the oxidation state of each S as +7, which exceeds the group maximum of +6. However, in peroxodisulfate there is an S–O–O–S (peroxo) group. In this, oxygen has the oxidation state −1. The oxidation state of S in peroxodisulfate is therefore +6.
(c) In thiosulfate, $S_2O_3^{2-}$, the average oxidation state is +2. However, there are clearly two different types of S in thiosulfate: the central S can be considered to be in the +6 oxidation state (thiosulfate is an analogue of sulfate itself), and the terminal S in the −2 oxidation state. The average of these is +2.

8.8   Polychalcogen Cations

There are many examples of polychalcogen cations. Oxygen, with its very limited tendency towards catenation, only forms O_2^+, but the other Group 16 elements, particularly sulfur and selenium, form a range of species, some of which are listed in Table 8.2. The structures of S_4^{2+} and S_8^{2+} are shown in Figure 8.10. These are prepared by oxidation of the elements using strong oxidizing agents, in non-aqueous solvents, for example equations 8.31 and 8.32.

In equation 8.31, AsF$_5$ is both an oxidizing agent (to oxidize the sulfur) and a source of the non-coordinating AsF$_6^-$ anion (by addition of F$^-$).

$$S_8 + 3AsF_5 \xrightarrow{\text{liquid SO}_2} S_8^{2+}(AsF_6^-)_2 + AsF_3 \quad (8.31)$$

$$6Te \xrightarrow{\text{oleum}} Te_6^{4+} \quad (8.32)$$

Table 8.2 Important polychalcogen cations

Element	Important cations
Oxygen	O_2^+
Sulfur	S_4^{2+}, S_8^{2+}, S_{19}^{2+}
Selenium	Se_4^{2+}, Se_8^{2+}, Se_{10}^{2+}
Tellurium	Te_4^{2+}, Te_6^{4+}

Consistent with the increasing metallic character of the Group 16 elements going down the group, tellurium forms the most highly oxidized cations (Te_6^{4+}), whereas sulfur forms both the largest and the least oxidized clusters, such as S_{19}^{2+}.

As with many other main group cluster ions, Wade's rules (see Sections 5.6.1 and 6.9) can be used to rationalize structures.

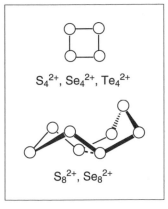

S_4^{2+}, Se_4^{2+}, Te_4^{2+}

S_8^{2+}, Se_8^{2+}

Figure 8.10 Structures of some polychalcogen cations

Worked Problem 8.7

Q Predict the shape of Se_4^{2+}.

A Se is in Group 16, and has 6 valence electrons. Each Se retains 2 electrons in a lone pair (Section 6.9), so the total electron count is $(4 \times 4) - 2$ (for the charge) = 14 electrons, *i.e.* 7 SEP. A six-vertex octahedron requires 7 SEP, but there are only 4 Se atoms, so there are two missing vertices. If the two opposing vertices of an octahedron are removed, a square results. Hence the structure of Se_4^{2+} is a square.

Summary of Key Points

1. *Variation in properties:* like Group 15, the elements of Group 16 cover the spectrum of properties from non-metallic oxygen to metallic polonium.

2. *Oxidation states:* oxygen is confined to a maximum possible oxidation state of +2 (in OF_2). The other elements form hypervalent compounds up to the group oxidation state of +6 (*e.g.* SO_3 and SF_6). The –2 state (which provides an octet of electrons) occurs for all elements.

3. *Catenation*, the ability of an element to form chains, is a commonly occurring theme in sulfur chemistry, second only to carbon.

4. *Strong pπ bonding* occurs between oxygen and many other elements. The oxides of the non-metallic elements are acidic (*e.g.* SO_3, P_4O_{10}), while those of metals are typically basic (*e.g.* Na_2O).

Problems

8.1. What is the oxidation state of the Group 16 element in the following compounds: (a) H_2S; (b) OF_2; (c) $H_2S_2O_7$ (pyrosulfuric acid); (d) S_2F_{10}; (e) hydrogen peroxide.

8.2. Identify the Group 16 element X in each of the following:
(a) X has a maximum oxidation state of +2.
(b) Oxyacids and oxyanions of X have a six-coordinate X atom.
(c) The oxides XO_2 and XO_3 are both polymeric at room temperature.
(d) The element forms a cation, X_2^+.
(e) All isotopes of X are radioactive.

8.3. Summarize the main types of compounds formed by sulfur which contain S–S bonds.

8.4. Describe how the following compounds might be prepared:
(a) anhydrous $ZnBr_2$, starting from $ZnBr_2.2H_2O$; (b) $PhC(O)OOH$;
(c) H_2S_3 starting from elemental sulfur.

8.5. Which of the following is not a stable substance: (a) AlS_2; (b) CaS; (c) BaS; (d) K_2S_2; (e) CS_2.

8.6. Predict the outcome of the following reactions:
(a) $AgNO_{3(aq)} + H_2S_{(g)} \rightarrow$
(b) $S_2O_3^{2-} + Br_2 \rightarrow$

8.7. Balance each of the following equations:
(a) $H_2S + O_2 \rightarrow SO_2 + H_2O$
(b) $H_2S + O_2F_2 \rightarrow SF_6 + HF + O_2$

8.8. Use $E°$ values to decide if the following reactions will proceed:
(a) $SO_2 + 2H_2S \rightarrow 3S + 2H_2O$
(b) $S_2O_3^{2-} + 2H^+ \rightarrow SO_2 + S + H_2O$ (disproportionation of thiosulfate in acid solution)

$S + 2H^+ + 2e^- \rightarrow H_2S$	$E°$ +0.14 V
$SO_2 + 4H^+ + 4e^- \rightarrow S + 2H_2O$	$E°$ +0.45 V
$S_2O_3^{2-} + 6H^+ + 4e^- \rightarrow 2S + 3H_2O$	$E°$ +0.5 V
$2SO_2 + 2H^+ + 4e^- \rightarrow S_2O_3^{2-} + H_2O$	$E°$ +0.4 V

References

1. K. I. Gobbato, M. F. Klapdor, D. Mootz, W. Poll, S. E. Ulic, H. Willner and H. Oberhammer, *Angew. Chem., Int. Ed. Engl.*, 1995, **34**, 2244.
2. M. Laing, *Educ. Chem.*, 1993, 140.

Further Reading

Stratospheric ozone depletion by chlorofluorocarbons (Nobel lecture), F. S. Rowland, *Angew. Chem., Int. Ed. Engl.*, 1996, **35**, 1786; *Polar ozone depletion (Nobel lecture)*, M. J. Molina, *Angew. Chem., Int. Ed. Engl.*, 1996, **35**, 1778.

The true allotropes of sulfur, G. Rayner-Canham and J. Kettle, *Educ. Chem.*, March 1991, 49.

Thiosulfate, S. W. Dhawale, *J. Chem. Educ.*, 1993, **70**, 12.

Developments in chalcogen-halide chemistry, B. Krebs and F.-P. Ahlers, *Adv. Inorg. Chem.*, 1990, **35**, 235.

9

The Group 17 (Halogen) Elements: Fluorine, Chlorine, Bromine, Iodine and Astatine

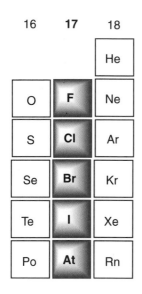

Aims

By the end of this chapter you should understand

- The unique features (small size, high electronegativity) of fluorine
- The decrease in electronegativity and increase in metallic character going down the group
- The use of redox potentials in rationalizing the reactions of halogen oxyanions and related species
- Pseudohalogens

9.1 Introduction and Oxidation State Survey

The halogens, like all of the p-block elements, span a range of reactivities. This ranges from the most electronegative and most reactive element in the Periodic Table (fluorine), to some of the least reactive, iodine and astatine.

The –1 oxidation state occurs in the halide anions (*e.g.* Cl⁻). This oxidation state becomes increasingly reducing on going down the group. Iodide, I⁻, is a moderate reducing agent, while chloride shows few reducing characteristics, except with very strong oxidizing agents. The trends in the stabilities of the main group halide compounds illustrate this, for example in Group 14. PbI_4 is a non-existent compound, owing to the combination of an oxidizing metal centre, Pb(IV), and a reducing iodide anion. In contrast, $PbCl_4$ and PbF_4 are more stable.

Positive oxidation states, +1, +3, +5 and +7, occur for chlorine, bromine and iodine, mainly in oxyanions and interhalogen compounds. Compounds in the highest oxidation states generally contain the electronegative elements oxygen and fluorine, *e.g.* IF_7 and IO_4^-. Fluorine, with its small size and high electronegativity, coupled with the weakness of the F–F bond, means that it is able to stabilize the very highest oxidation states of elements, *e.g.* AuF_5, NiF_4, PtF_6.

Iodine (and to a lesser extent bromine) forms solvated cations (with iodine in the +1 oxidation state) such as $[I(pyridine)_2]^+$, by reaction of I_2, pyridine and $AgNO_3$ in a non-aqueous solvent (equation 9.1).

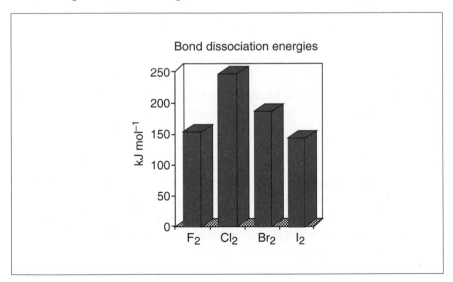

The structure of the $[I(pyridine)_2]^+$ cation

$$I_2 + AgNO_3 + 2 \text{ pyridine} \rightarrow [I(pyridine)_2]^+NO_3^- + AgI \quad (9.1)$$

9.2 The Elements

All the elements are reasonably abundant in Nature, but always in compounds. In pure form they exist as diatomic molecules, X_2. At room temperature, F_2 and Cl_2 are both gases (yellow and pale green, respectively), Br_2 is a deep red liquid, while I_2 is a metallic purple solid, showing the increasing intermolecular (van der Waals) forces for the heavier halogens. For chlorine, bromine and iodine the bond dissociation energies decrease going down the group (Figure 9.1), owing to poorer overlap between increasingly large atoms with diffuse orbitals. Fluorine has an anomalously low bond dissociation energy, because of increased repulsion of lone pairs on adjacent fluorine atoms, owing to their closer proximity in F_2 compared to Cl_2 (see also Worked Problem 7.4).

Bond dissociation energies

y-axis: kJ mol^{-1} (0, 50, 100, 150, 200, 250); x-axis: F_2, Cl_2, Br_2, I_2

Figure 9.1 Bond dissociation energies of the halogens X_2

9.2.1 Fluorine

Fluorine occurs in the mineral *fluorspar*, CaF_2, and also *fluorapatite*, $CaF_2.3Ca_3(PO_4)_2$. Elemental fluorine is obtained by the electrolysis of

Fluorine is used to manufacture non-stick polymers [such as poly(tetrafluoroethene)] and fluorocarbon refrigerants.

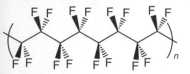

Poly(tetrafluoroethene) PTFE

potassium fluoride dissolved in anhydrous hydrogen fluoride (equation 9.2).

$$2HF + 2KF \rightarrow 2KHF_2 \rightarrow 2KF + H_2 + F_2 \qquad (9.2)$$

9.2.2 Chlorine

Chlorine occurs as sodium chloride in large underground salt deposits (formed by the evaporation of ancient seas), from which it is obtained commercially. Alternatively, sea water contains a relatively high chloride concentration (15,000 ppm). Elemental chlorine is obtained by the electrolysis of *aqueous* salt solutions, which produces sodium hydroxide as a by-product. Chlorine compounds have numerous industrial and household applications.

9.2.3 Bromine

Bromide ions occur in seawater (concentration 30 ppm), from which Br_2 is obtained by reaction with chlorine (equation 9.3).

$$Cl_2 + 2Br^- \rightarrow 2Cl^- + Br_2 \qquad (9.3)$$

9.2.4 Iodine

Iodine is found naturally as sodium iodate ($NaIO_3$), from which elemental iodine is obtained by reduction. Iodine ions are found in some brines, from which the element is obtained by oxidation with chlorine (compare equation 9.3). Iodine begins to show some properties reminiscent of metals, such as the lustrous semi-metallic appearance of the element (and high electrical conductivity under pressure).

9.2.5 Astatine

The longest-lived astatine isotope, ^{210}At, has a half-life of 8.3 h.

All isotopes of astatine are radioactive, and trace amounts occur in uranium ores. More usually, astatine isotopes are manufactured by the irradiation of a bismuth target with α particles, from which elemental astatine (possibly in the At_2 form) is obtained in the form of the isotopes ^{209}At, ^{210}At and ^{211}At.

9.3   Chemistry of the Elements

All of the elemental halogens are oxidizing agents, with the reactivity decreasing going down the group. Elemental fluorine is the most reactive of any element, and spontaneously forms chemical compounds with all other elements except the lighter noble gases helium, neon and argon. Reactions with fluorine must be carried out in special vessels, such as poly(tetrafluoroethene) (PTFE), in metals such as nickel which form a passivating layer of the metal fluoride, or in very dry glass vessels. Chlorine and bromine are far less reactive than fluorine, but they still react with many elements directly, while iodine is the least reactive of the four, and it often requires heating for reaction to proceed.

9.4 Hydrogen Halides, HX, and Halide Salts, X⁻

Box 9.1 Laboratory Preparations of Gaseous HX:

$$CaF_2 + \text{concentrated } H_2SO_4 \rightarrow CaSO_4 + 2HF \quad (9.4)$$
$$NaCl + \text{concentrated } H_2SO_4 \rightarrow NaHSO_4 + HCl \quad (9.5)$$
$$2P_{(red)} + 3X_2 + 6H_2O \rightarrow 2H_3PO_3 + 6HX \ (X = Br, I) \quad (9.6)$$

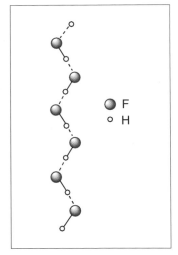

All hydrogen halides are known, and stability decreases down the group, owing to the increasing mismatch in atomic sizes. HI is rather unstable (the $\Delta_f H°$ is positive, $+26.5$ kJ mol⁻¹), and HI exists in equilibrium with H_2 and I_2. The melting and boiling points increase going down the group, with the exception of HF which has an anomalously high boiling point due to hydrogen bonding (compare Figure 2.4). This hydrogen bonding also occurs in the gas phase, where HF vapour mainly exists as a hexamer, $(HF)_6$, up to 60 °C. In the solid state, HF exists in a polymeric zigzag chain, $(HF)_n$, with strong F···H–F hydrogen bonds (Figure 9.2). The F···H–F unit is always linear about the hydrogen, and also occurs in so-called hydrogenfluoride salts such as $K^+HF_2^-$.

The hydrogen halides dissolve in water to give highly acidic solutions (equation 9.7).

$$HX + H_2O \rightleftharpoons H_3O^+_{(aq)} + X^-_{(aq)} \quad (9.7)$$

Figure 9.2 The structure of HF in the solid state

HCl, HBr and HI are all very strong acids and are essentially fully dissociated in dilute solution; however, HF is a much weaker acid. Aqueous HF solutions contain undissociated HF, together with fragments of the $(HF)_n$ polymeric chain such as $H_2F_3^-$ and $H_3F_4^-$, held together by the strong H···F hydrogen bonding that is so important in this system.

Some important pseudohalides:
Azide N_3^-
Cyanide CN^-
Cyanate NCO^-
Thiocyanate NCS^-

Box 9.2 Pseudohalogens

A number of monoanions are known which behave in a manner similar to halides, and form similar compounds: these anions (X^-) are called **pseudohalides**. The neutral dimers (X_2) are **pseudohalogens**, though not all pseudohalides form pseudohalogens; for example, azide (N_3^-) is a well-known pseudohalide, but the dimer ($N_3)_2$ is unknown. The **inter-pseudohalogen** iodine azide $I-N_3$ is explosive.

Worked Problem 9.1

Q Would you expect Ag^+ and Pb^{2+} salts of the azide ion to be soluble in water?

A The azide is a pseudohalogen, and since silver halides and lead(II) halides are insoluble in cold water, the compounds AgN_3 and $Pb(N_3)_2$ are also expected to be insoluble.

AgN_3 and $Pb(N_3)_2$, like other heavy metal azide salts, are highly explosive.

9.5   Halogen Oxides

All of the halogens form oxides; those of iodine are the most stable, while those of bromine tend to be the least stable; all are oxidizing agents.

9.5.1 Dihalogen Monoxides, E_2O

Fluorine, chlorine and bromine form compounds of the type E_2O. OF_2 is the most stable, and is prepared by reaction of elemental fluorine with dilute aqueous sodium hydroxide (equation 9.8).

$$2F_2 + 2OH^- \rightarrow OF_2 + 2F^- + H_2O \qquad (9.8)$$

However, OF_2 is a powerful oxidizing agent, and will oxidize water to oxygen (equation 9.9). It explodes when mixed with halogens, but a spark is required to ignite mixtures of OF_2 and hydrogen or CO.

$$OF_2 + H_2O \rightarrow O_2 + 2HF \qquad (9.9)$$

Since the order of decreasing electronegativity is F > O > Cl, we write formulae as OF_2 and Cl_2O, with the *least* electronegative element first.

Cl_2O and Br_2O are prepared by the reaction of chlorine or bromine with mercury(II) oxide (equation 9.10).

$$2Cl_2 + 2HgO \rightarrow Cl_2O + HgCl_2.HgO \qquad (9.10)$$

9.5.2 Chlorine Dioxide, ClO_2

ClO_2 is prepared as a yellow gas, explosive in high concentrations, by the reduction of potassium chlorate ($KClO_3$) in acidic solution. ClO_2 is a paramagnetic molecule with a bent shape (Figure 9.3a); it shows little tendency to dimerize, because the unpaired electron is delocalized over the whole molecule. However, in the solid state it forms very loose dimers, with a Cl···O distance of 270.8 pm compared to the Cl–O bond lengths of around 147 pm (Figure 9.3b).[1]

Chlorine dioxide is commercially important as a bleaching and sterilizing agent.

(a)

147 pm

117.6°

Cl

O O

(b) 147.6 pm   270.8 pm

147 pm

115.6°

Figure 9.3 The structure of ClO_2 in (a) the gas phase and (b) the solid phase

9.5.3 Other Chlorine and Bromine Oxides

The oxides of chlorine (and to a lesser extent bromine) show some structural similarities in that many of them contain halogen–oxygen–halogen linkages, where the halogen atoms can be in various oxidation states: +1, +3, +5 and +7.

The structures of selected chlorine and bromine oxides of this type are shown in Figure 9.4. Of these, Cl_2O_7 is the most stable, and is the anhydride of chloric(VII) acid, $HClO_4$. Cl_2O_6 ionizes to $ClO_2^+ClO_4^-$ in the solid state. The monomer ClO_3 is formed by pyrolysis of chlorine chlorate(VII), $ClOClO_3$ (Cl_2O_4 in Figure 9.4).[2]

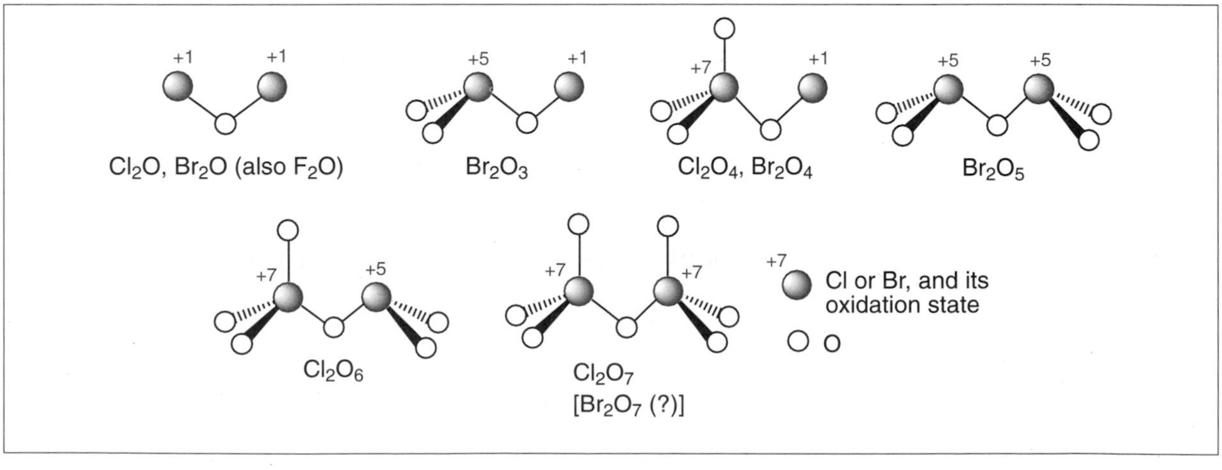

Figure 9.4 Chlorine and bromine oxides containing two halogen atoms

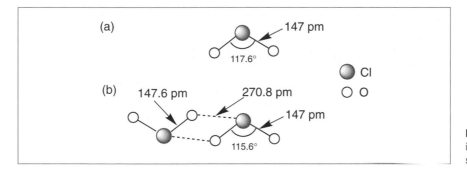

+1   +1   +5   +1   +7   +1   +5   +5

Cl_2O, Br_2O (also F_2O) Br_2O_3 Cl_2O_4, Br_2O_4 Br_2O_5

+7   +5   +7   +7   +7

Cl_2O_6 Cl_2O_7 Cl or Br, and its
                   [Br_2O_7 (?)]   oxidation state

O O

The bromine oxides are less stable than the chlorine analogues, and stable compounds are constructed from –O–Br and –O–BrO$_2$ units. Isomers may occur; for example, in materials of stoichiometry BrO_2 there may be O_2Br–BrO_2, OBr–O–BrO_2 and Br–O–BrO_3 ['bromine bromate(VII)']; Br_2O_4 is shown in Figure 9.4.[3]

9.5.4 Iodine Oxides

Of all the halogen oxides, the ones of iodine, particularly in higher oxidation states, are the most stable. Iodine(V) oxide, I_2O_5, which decomposes only at temperatures greater than 300 °C, is made by heating HIO_3, of which it is the anhydride (equation 9.11). Iodine oxides, unlike the chlorine and bromine ones, have polymeric structures.

$$2HIO_3 \rightleftharpoons I_2O_5 + H_2O \qquad (9.11)$$

9.6   Oxyacids and Oxyanions of the Halogens

Acid	Acid dissociation constant, K_a
HOCl	2.8×10^{-8}
HClO$_2$	1.0×10^{-2}
HClO$_3$	10^3
HClO$_4$	10^7

There are well-established series of oxyacids and their conjugate anions for each halogen, though not all halogens form all types of species. Structures of the different types of acids are shown in Figure 9.5. For a given element, the strength of the acid increases with oxidation state.

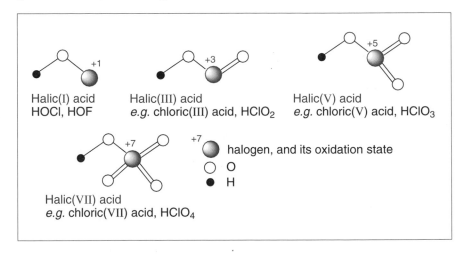

Figure 9.5 Oxidation states and nomenclature of halogen oxyacids

Halic(I) acid
HOCl, HOF

Halic(III) acid
e.g. chloric(III) acid, HClO$_2$

Halic(V) acid
e.g. chloric(V) acid, HClO$_3$

Halic(VII) acid
e.g. chloric(VII) acid, HClO$_4$

halogen, and its oxidation state
O
H

9.6.1 Halic(I) Acids (HOX) and Halate(I) Salts (OX$^-$)

The only oxyacid of fluorine which is known is fluoric(I) acid, HOF, obtained by the fluorination of ice at low temperatures (equation 9.12).

$$F_2 + H_2O \rightarrow HF + HOF \qquad (9.12)$$

The other halogens show the same general behaviour, though the stability of the HOX formed decreases dramatically on going from F to Cl to Br to I. Halic(I) anions can be obtained by reaction of the halogen with hydroxide solutions, and for chlorine the equilibrium lies far to the right-hand side (equation 9.13).

$$X_2 + 2OH^- \rightleftharpoons XO^- + X^- + H_2O \qquad (9.13)$$

Equilibrium constants (K) for equation 9.13:

X	K
Cl	7×10^{15}
Br	2×10^8
I	30

However, the halic(I) anions tend to undergo **disproportionation** reactions (see Section 6.5.2), to give halide and halate(V) (XO_3^-) anions (equation 9.14).

$$3XO^- \rightleftharpoons XO_3^- + 2X^- \qquad (9.14)$$

i.e.

$$\text{halate(I)} \rightleftharpoons \text{halate(V)} + \text{halide}$$

Equilibrium constants (K) for equation 9.14:

X	K
Cl	10^{27}
Br	10^{15}
I	10^{20}

Since the equilibria in reactions 9.13 and 9.14 lie to the right-hand side, the actual products formed are dependent on the relative **rates of reactions**. When chlorine is dissolved in hydroxide solution, ClO^- is formed, but the rate of disproportionation at room temperature to ClO_3^- and Cl^- (equation 9.14) is very slow (ClO_3^- can be obtained by passing Cl_2 into *hot* OH^- solution). For bromine, both reactions 9.13 and 9.14 are fast, so a mixture of BrO^-, BrO_3^- and Br^- is obtained (the proportions being temperature dependent), while for iodine, reaction 9.14 is fast, and so I^- and IO_3^- are essentially the only products formed.

9.6.2 Halic(III) Acids (HXO$_2$) and Halate(III) Salts (XO$_2^-$)

Chloric(III) acid ($HClO_2$) is the only free acid definitely established (although $HBrO_2$ may exist), but it is unstable. In contrast, chlorates(III) (ClO_2^-) and bromates(III) (BrO_2^-) are well known.[4] $HClO_2$ is formed by acidification of the chlorate(III) anion, which in turn is formed by reaction of ClO_2 with hydroxide (equation 9.15).

$$2ClO_2 + 2OH^- \rightarrow ClO_2^- + ClO_3^- + H_2O \qquad (9.15)$$

9.6.3 Halic(V) Acids (HXO$_3$) and Halate(V) Salts (XO$_3^-$)

Halic(V) acids can be isolated for Cl, Br and I, with iodic(V) acid (HIO_3) being the most stable and the only one which can be isolated as a solid. The halic(V) acids are weaker oxidizing agents and stronger acids than the corresponding halic(III) acids; refer to Box 9.3 for E° values.

9.6.4 Halic(VII) Acids (HXO$_4$) and Halate(VII) Salts (XO$_4^-$)

Chlorate(VII) salts are often referred to as **perchlorates**.

Halic(VII) acids of chlorine, bromine and iodine are known. Chloric(VII) acid is a strong oxidizing agent, but the least strongly oxidizing of all chlorine oxyacids. The chlorate(VII) anion can be used to crystallize large cations. However, there are numerous reports, anecdotes and warnings on the explosive nature of chlorate(VII) salts of oxidizable cations.

Iodic(VII) acid (HIO$_4$) and iodate(VII) salts, containing the tetrahedral IO$_4^-$ anion, exist. However, the chemistry of iodine(VII) oxyacids and oxyanions is dominated by compounds with higher coordination numbers for iodine. This occurs by addition of water across two I=O bonds to give two I–OH groups, and formation of I–O–I bridges. Iodine therefore shows similarities to the other heavy p-block elements tellurium (see Section 8.7.2) and bismuth (see Section 7.4.3) in this aspect of its chemistry. Thus, acidic, aqueous iodate(VII) solutions contain a variety of species including H$_5$IO$_6$ and various deprotonated forms of this acid. In strongly acidic solution the I=O group of H$_5$IO$_6$ can be protonated to give I(OH)$_6^+$. H$_5$IO$_6$ has two of the five protons readily exchangeable. Heating H$_5$IO$_6$ causes loss of water in two stages (equation 9.16).

$$H_5IO_6 \xrightarrow{80°C} H_4I_2O_9 \xrightarrow{100°C} HIO_4 \qquad (9.16)$$

Salts containing the [I$_2$O$_9$]$^{4-}$ ion are known, and contain two face-sharing IO$_6$ octahedra, while in the related ion [H$_2$I$_2$O$_{10}$]$^{4-}$ there are two edge-sharing IO$_6$ octahedra.

In contrast to chlorates(VII) and iodates(VII), which have been known for many years, bromic(VII) acid eluded synthesis until 1968. This 'middle element anomaly' manifests itself on numerous occasions in the chemistry of As, Sb and Br, because these elements (which form after the poorly screening 3d shell has been filled) are more difficult to oxidize than expected. Bromate(VII) was eventually obtained by oxidation of bromate(V) in alkaline solution by F$_2$ (equation 9.17), and bromic(VII) acid was obtained from it by protonation. The salts exist solely as the tetrahedral BrO$_4^-$ ion, with no evidence for higher coordination number adducts as with iodic(VII) acid. The bromate(VII) ion is a strong oxidizing agent.

$$BrO_3^- + F_2 + 2OH^- \rightarrow BrO_4^- + 2F^- + H_2O \qquad (9.17)$$

Structure of H$_5$IO$_6$ = I(O)(OH)$_5$
178 pm, 189 pm

Structure of the I$_2$O$_9^{4-}$ anion

Structure of the H$_2$I$_2$O$_{10}^{4-}$ anion

Box 9.3 Using Redox Potentials to Rationalize the Chemistry of the Halogens

The halogens form a diverse range of oxy species, and standard redox ($E°$) potentials are very useful in rationalizing their reactivities. The $E°$ values for chlorine species in acid and alkaline solution are given in Figure 9.6.

- The ClO_4^- ion and $HClO_4$ (containing chlorine in the +7 oxidation state) can only act as an oxidizing agent.
- Species in lower oxidation states (except Cl^-) can act as reducing agents, but they more generally act as oxidizing agents, *e.g.* ClO_3^- is able to oxidize Br^- ions:

$$ClO_3^- + 6H^+ + 6e^- \rightarrow Cl^- + 3H_2O \qquad E° \;+1.442 \text{ V}$$
$$6Br^- \rightarrow 3Br_2 + 6e^- \qquad E \;-1.077 \text{ V}$$

Thus:

$$ClO_3^- + 6H^+ + 6Br^- \rightarrow Cl^- + 3Br_2 + 3H_2O \qquad E° \;+0.365 \text{ V}$$

The value of $E°$ is positive, so the reaction will proceed under standard conditions.

- The oxyanions are much stronger oxidizing agents (shown by the more positive $E°$ values) in acidic than in basic solution. The exception is for $Cl_2 \rightarrow 2Cl^-$, since neither H^+ nor OH^- appear in the half-equations.
- Many species are susceptible to disproportionation, *e.g.* $HClO_2$:

$$HClO_2 + 2H^+ + 2e^- \rightarrow HClO + H_2O \qquad E° \;+1.673 \text{ V}$$
$$HClO_2 + H_2O \rightarrow ClO_3^- + 3H^+ + 2e^- \qquad E \;-1.157 \text{ (reverse sign)}$$

Thus:

$$2HClO_2 \rightarrow HClO + ClO_3^- + H^+ \qquad E° \;+0.516 \text{ V}$$

Since the $E°$ is positive, $HClO_2$ is unstable in solution.

For the line joining $HClO_2$ and $HClO$ in acid solution:
$$HClO_{2(aq)} + 2H^+_{(aq)} + 2e^- \rightarrow HClO_{(aq)} + H_2O$$
$$E° = + 1.673 \text{ V}$$

9.7   Interhalogen Compounds

9.7.1 Neutral Compounds

There is a wide range of interhalogen compounds (containing two or more different halogens), with many examples of neutral, cationic and anionic compounds. The simplest are the heteronuclear diatomic molecules X–Y, all of which are known. These are formed by simply mixing the two component halogens, *e.g.* for IBr (equation 9.18).

Heteronuclear diatomic interhalogens and their enthalpies of formation:

	$\Delta_f H°$ (kJ mol^{-1})
IF	−94.5
BrF	−58.5
ClF	−53.6
ICl	−35.3
IBr	−10.5
BrCl	+14.6

Note that the more electropositive halogen is written first.

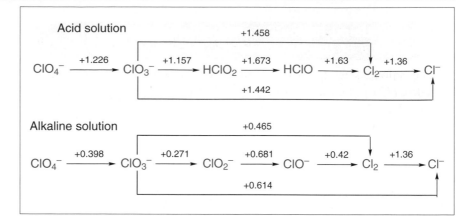

Figure 9.6 Values of $E°$ (V) for chlorine-containing species in acid and alkaline solution

---

Worked Problem 9.2

Q Show that the disproportionation of chlorine, to chloride and chlorate(I), is favoured in basic solution.

A The two half-equations are:

$$Cl_2 + 2e^- \rightarrow 2Cl^- \qquad\qquad E° +1.36 \text{ V}$$
$$Cl_2 + 4OH^- \rightarrow 2ClO^- + 2H_2O + 2e^- \quad E -0.42 \text{ V}$$

Adding:

$$Cl_2 + 2OH^- \rightarrow ClO^- + Cl^- + H_2O \qquad E +0.94 \text{ V}$$

With a positive E, the reaction will proceed, under standard conditions.

$$I_2 + Br_2 \rightarrow 2IBr \qquad\qquad (9.18)$$

Thus, IF is the most stable with respect to the parent I_2 and F_2, because of a dipolar contribution $I^{\delta+}$–$F^{\delta-}$ to the bonding.

Worked Problem 9.3

Q IF readily undergoes disproportionation into I_2 and IF_5. Write a balanced equation for the reaction and, using the following $\Delta_f H°$ values, show that the reaction is thermodynamically favoured: $\Delta_f H°(IF_5)$ –822 kJ mol^{-1}, $\Delta_f H°(IF)$ –94.5 kJ mol^{-1}.

A

$$10IF \rightarrow 4I_2 + 2IF_5 \qquad (9.19)$$

ΔH for this reaction $= 2\Delta_f H°(IF_5) - 10\Delta_f H°(IF) = 2(-822) - 10(-94.5) = -699$ kJ mol^{-1}, so the reaction is favoured.

Other neutral interhalogens include compounds of the type EF_3, EF_5 (known where E = Cl, Br and I, though some are unstable) and, for iodine, IF_7. Direct combination of the elements provides a general method for the synthesis of many of this type of compound.

Organic derivatives such as $MeIF_2$ and $PhICl_2$ are well-known compounds. $PhICl_2$, a crystalline solid, is sometimes used as a conveniently prepared and handled oxidizing agent, formed by passing excess Cl_2 through a trichloromethane solution of iodobenzene (equation 9.20). Indeed, all of the interhalogens are strong oxidizing agents, the most reactive being ClF_3 which is as powerful as elemental fluorine.

$$PhI + Cl_2 \rightarrow PhICl_{2(s)} \qquad (9.20)$$

The shapes of all of these interhalogen molecules are readily predicted by VSEPR (see Section 1.4). Selected examples are illustrated and covered in the worked examples and problems.

ClF$_3$

T-shaped

BrF$_5$

Square pyramidal

Worked Problem 9.4

Q The only known higher interhalogen chloride is ICl_3, which crystallizes as a dimer, I_2Cl_6. Predict the shape of I_2Cl_6 using VSEPR.

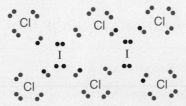

Lewis dot diagram for I_2Cl_6

A Lewis structure:
I has 7 valence electrons, plus 3 from the Cl atoms, giving 10 electrons, or 5 pairs. A dative bond from Cl→I donates another pair,

giving 6 pairs at each I, in an octahedral arrangement. Each I has two *trans* lone pairs, and the molecule is therefore planar overall (Figure 9.7).

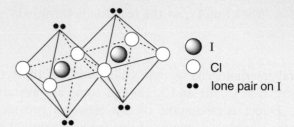

Figure 9.7 The structure of I_2Cl_6

9.7.2 Interhalogen Cations and Anions

Many interhalogen cations and anions are also known; a selection is given in Table 9.1. The fluoride compounds are generally formed by addition or removal of F^- from a parent neutral compound (*e.g.* ClF_3 in equations 9.21 and 9.22), similar to the formation of the related noble gas compounds (see Chapter 10), with which the interhalogens share much common chemistry. For example:

Fluoride addition: $ClF_3 + Cs^+F^- \rightarrow Cs^+ClF_4^-$ (9.21)

Fluoride removal: $ClF_3 + AsF_5 \rightarrow ClF_2^+AsF_6^-$ (9.22)

Oxidation and fluoride ion removal is equivalent to the addition of F^+, which is not stable on its own.

Alternatively, vigorous oxidation (using KrF_2, see Section 10.3.1) of a lower oxidation state compound can be used (equation 9.23).

$$ClF_5 + KrF_2 + AsF_5 \rightarrow ClF_6^+AsF_6^- + Kr   (9.23)$$

Table 9.1 Selected interhalogen cations and anions

Oxidation state	+1	+3	+5	+7
Cations	ClF_2^+	ClF_4^+	ClF_6^+	
	ICl_2^+	BrF_4^+	BrF_6^+	
		IF_4^+	IF_6^+	
Anions	$BrCl_2^-$	ClF_4^-	BrF_6^-	
	ICl_2^-	BrF_4^-	IF_6^-	IF_8^-
	IBr_2^-	ICl_4^-		
	I_3^-			

In this case, the compound ClF_7 does not exist (seven fluorines cannot fit around a Cl atom), but by forming a six-coordinate ClF_6^+ cation the chlorine(VII) oxidation state is stabilized.

Related to the interhalogen compounds are mixed oxide–halide species, of the type IOF_5, $[IOF_6]^-$, *etc*. Many of these species have analogous xenon compounds.

Worked Problem 9.5

Q How would you synthesize the following compounds, starting from the binary fluoride, plus any other reagents of your choice: (a) $[NO]^+[IF_8]^-$; (b) $[ClF_2]^+[RuF_6]^-$.

A (a) $[IF_8]^-$ is formed by the addition of F^- to IF_7. Hence the source of fluoride should be from NOF (see Section 7.4.1):

$$NOF + IF_7 \rightarrow [NO]^+[IF_8]^-$$

(b) $[ClF_2]^+$ can be formed by the loss of F^- from ClF_3. RuF_5 is the appropriate fluoride-abstracting reagent:

$$ClF_3 + RuF_5 \rightarrow [ClF_2]^+[RuF_6]^-$$

Worked Problem 9.6

Q Discuss the observation that ClF_6^+ and ClF_6^- both have octahedral geometries.

A ClF_6^+ is a six electron-pair species (in the Cl valence shell), and so by VSEPR (see Section 1.4) it is expected to have an octahedral shape. ClF_6^- has seven electron-pairs; it must therefore have a stereochemically inactive lone pair of electrons residing in an s-orbital.

9.7.3 Polyiodide Anions

Polyiodide anions are one type of interhalogen species commonly encountered in the laboratory and are prepared from mixtures of iodine and a suitable iodide salt, in appropriate proportions. The tri-iodide anion, I_3^-, is the best known (equation 9.24).

$$I_2 + I^- \rightleftharpoons I_3^- \tag{9.24}$$

The I_3^- anion is most commonly encountered during redox titrations involving iodine, where an excess of iodide ions is added to convert any iodine formed (which is only sparingly soluble in water) to the soluble I_3^- ion. Indeed, equation 9.25, which is very widely used in high school and undergraduate chemistry, is more correctly represented as 9.26.

$$I_2 + 2S_2O_3^{2-} \rightarrow 2I^- + S_4O_6^{2-} \tag{9.25}$$

$$I_3^- + 2S_2O_3^{2-} \rightarrow 3I^- + S_4O_6^{2-} \tag{9.26}$$

The solubility of a tri-iodide salt $M^+I_3^-$ is dependent on the size of the cation M^+, and to precipitate the I_3^- ion from solution, large cations such as Me_4N^+ or Ph_4As^+ are often used. The cation also has a marked influence on the *structure* of the anion, as shown in Figure 9.8; the smaller Cs^+ ion distorts the I_3^- ion, whereas the larger Ph_4As^+ does not.

Recall from Section 3.5 that a large cation stabilizes a large anion.

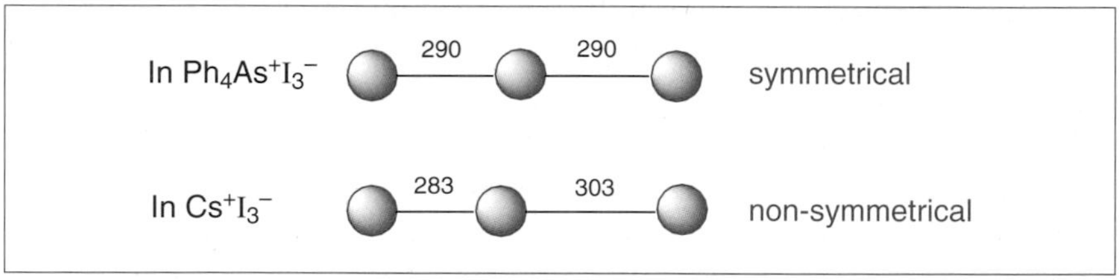

Figure 9.8 Structures of the I_3^- anions in two different salts, with bond lengths (pm)

Longer **polyiodide** anions, such as I_5^-, I_7^- and I_9^-, are also known *in the solid state*, formed by addition of iodide to two, three and four I_2 molecules, respectively.

Worked Problem 9.7

Q To synthesize the compound $[Ph_4As][I_7]$, what compounds, and in what molar proportions, should be reacted together?

A The starting source of the $[Ph_4As]^+$ ion should be $[Ph_4As]I$, to form the I_7^- ion; another $3I_2$ (giving 6 I atoms) are needed. Hence $[Ph_4As]I$ and I_2 in a 1:3 molar ratio should be used.

9.8   The Chemistry of Astatine

Owing to the radioactivity and limited availability of astatine, the chemistry of this element is done using tracer studies. The chemistry can be followed by adding trace amounts of astatine to iodine and then following the chemistry of astatine by monitoring the radioactivity.

Like iodine, astatine forms compounds in at least four oxidation states (−1, 0, +1, +5 and possibly +3), although unlike iodine it does not appear to form compounds in the +7 oxidation state. In the −1 state the At^- ion is formed, and in the higher oxidation states the AtO^-, AtO_2^- and AtO_3^- ions appear to exist. The elemental form, which may be of the form At_2 (although this has not been proven), is volatile, soluble in non-polar organic solvents (such as CCl_4) and is reduced to At^- using reducing agents such as SO_2.

The properties of At are as expected for the element's position in the group, based on extrapolation of the properties for F, Cl, Br and I.

Worked Problem 9.8

Q If you were investigating the chemistry of astatine by tracer studies (with iodine) and using elemental astatine as a starting compound, which reagents would you use to make and confirm the existence of AtO_3^-?

A The elemental astatine would be diluted with iodine (I_2) and oxidized with a strong oxidizing agent (such as acidified $NaClO_3$). Under the same conditions, iodine gives iodate, IO_3^-. Since barium iodate is insoluble in water, addition of a soluble barium salt (for example, barium nitrate) to the iodate solution would give a precipitate of $Ba(IO_3)_2$. If astatine was behaving similarly, the radioactivity would be present in the precipitate.

Summary of Key Points

1. The *dominant features of the chemistry of the halogens* are the formation of X^- anions, or covalent compounds with strong M–X bonds.

2. *Many interhalogen species are known*, as neutral, cationic and anionic species, and are most stable when the central atom is iodine, in combination with fluorine.

3. *Fluorine is an extremely reactive element*, able to stabilize the highest oxidation states in compounds with other elements.

4. *A wide range of oxyacids and oxyanions are formed*, except for fluorine; the compounds are all strong oxidizing agents.

Problems

9.1. What is the oxidation state of the Group 17 element(s) in the following compounds or ions: (a) $HClO_4$; (b) ClF_4^-; (c) Cl_2O; (d) $HClO_3$; (e) hypochlorite anion.

9.2. Identify the element X in each of the following:
(a) X has the strongest X–X bonds of any element in the group.
(b) X forms oxyanions XO_n^- where n is 1, 2, 3 or 4, and fluorides XF_n where n is 1, 3, 5 or 7.
(c) X is the most electronegative of all the elements.
(d) X forms fewer, and less stable, oxides than the elements above it and below it.

9.3. What happens to the values of the following properties of Group 17 elements, in the sequence Cl, Br, I? Give reasons. (a) Enthalpy of vaporization; (b) bond energy of the X_2 molecule; (c) first ionization energy; (d) atomic radius.

9.4. Balance the following equations:
(a) $Cl_2 + IO_3^- \rightarrow IO_4^-$ in alkaline solution
(b) $KMnO_4 + KCl + H_2SO_4 \rightarrow MnSO_4 + K_2SO_4 + H_2O + Cl_2$

9.5. Liquid BrF_5 and liquid AsF_5 are both poor conductors of electricity, but a liquid mixture of the two is a much better conductor. Explain why.

9.6. Which of the following halides are rapidly hydrolysed: BCl_3, $SiCl_4$, CCl_4, SF_6, PCl_5?

9.7. How would you synthesize the following compounds: (a) $Cs^+[ClF_4]^-$; (b) $Cs^+[BrF_6]^-$.

9.8. Cl_2O_6 ionizes to $ClO_2^+ClO_4^-$ in the solid state. Find (from the other chapters in this book) other examples of molecular gas-phase species which crystallize as ionic solids.

9.9. What reagents would you use to make (from elemental astatine) and confirm the existence of the At^- ion?

9.10. Using VSEPR (see Chapter 1) show that the I_3^- ion is predicted to be linear.

9.11. The cyanide ion (CN⁻) is a *pseudohalogen* (see Box 9.2) and has many properties in common with halides (*e.g.* Cl⁻); for example, it forms cyanogen $(CN)_2$, which is analogous to Cl_2. Based on this assumption, predict the outcome of the following reactions:
(a) $(CN)_2$ + excess aqueous NaOH →
(b) $Ag^+_{(aq)}$ + $CN^-_{(aq)}$ →
(c) I_2 + $(CN)_2$ →
(d) $(CN)_2$ + H_2 →
(e) $(CN)_2$ + reducing agent →

9.12. Show using $E°$ values (Figure 9.6) that Cl_2 is susceptible to disproportionation in basic solution, but not in acid solution, under standard conditions.

References

1. A. Rehr and M. Jansen, *Angew. Chem., Int. Ed. Engl.*, 1991, **30**, 1510; A. Rehr and M. Jansen, *Inorg. Chem.*, 1992, **31**, 4740.
2. H. Grothe and H. Willner, *Angew. Chem., Int. Ed. Engl.*, 1994, **33**, 1482.
3. T. R. Gilson, W. Levason, J. S. Ogden, M. D. Spicer and N. A. Young, *J. Am. Chem. Soc.*, 1992, **114**, 5469; D. Leopold and K. Seppelt, *Angew. Chem., Int. Ed. Engl.*, 1994, **33**, 975.
4. W. Levason, J. S. Ogden, M. D. Spicer, M. Webster and N. A. Young, *J. Am. Chem. Soc.*, 1989, **111**, 6210.

Further Reading

Bromine oxides, K. Seppelt, *Acc. Chem. Res.*, 1997, **30**, 111.

10

The Group 18 (Noble Gas) Elements: Helium, Neon, Argon, Krypton, Xenon and Radon

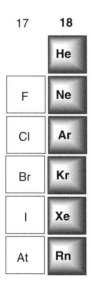

The **noble gases** have also been called the **rare gases**, and the **inert gases**. None of the names is particularly suitable; with extensive chemistry, xenon is not inert, and argon is almost 30 times more abundant than carbon dioxide in air.

Aims

By the end of this chapter you should understand:

- The greater reactivity of xenon compared to the other elements
- The stability of xenon compounds containing electronegative elements (O and F)
- The reactivity of noble gas fluorides as fluoride ion donors and acceptors

10.1 Introduction and Oxidation State Survey

The noble gases are characterized by extremely high ionization energies (Table 10.1), and for a long time it was thought that chemical compounds of these elements could not be formed. The first xenon compounds were synthesized in 1962, and since then the chemistry of xenon has become reasonably extensive; a few compounds of krypton have also been reported. There are several summaries of the early historical developments in noble gas chemistry.[1]

Many of the noble gas compounds have analogues in the chemistry of the heavier Group 17 elements, particularly iodine. VSEPR (see Section 1.4) is a powerful technique for rationalizing and predicting the shapes of many noble gas compounds, which have relatively large numbers of electron pairs in the valence shell of the central atom.

Noble gas atoms have filled shells, and so in order to form chemical compounds, electrons must be promoted into the next shell. Depending on the number of electrons promoted, compounds of the noble gases in oxidation states +2, +4, +6 and +8 have all been synthesized. With

Table 10.1 Atomic masses, first ionization energies (IEs), boiling points and van der Waals radii of the noble gases

	Atomic mass (g mol^{-1})	First IE (kJ mol^{-1})	Boiling point (K)	van der Waals radius[a] (pm)
He	4.0	2379	4.2	–
Ne	20.2	2087	27.1	157.8
Ar	39.9	1527	87.3	187.4
Kr	83.8	1357	121.3	199.6
Xe	131.3	1177	166.1	216.7
Rn	222.0	1043	208.2	–

[a]Values for the noble gases in the solid state from: S. S. Batsanov, *J. Chem. Soc., Dalton Trans.*, 1998, 1541.

increasing atomic size, the outer electrons are more easily removed and there is a decrease in first ionization energy, so there are many xenon compounds but few for krypton, and none have been isolated for argon. All compounds tend to be strong oxidizing agents, and the fluorides are also powerful fluorinating agents. The formation of *anionic* noble gas compounds (such as caesium neonide, Cs^+Ne^-) with electropositive metals is an intriguing possibility, though none has yet been isolated.[2]

Worked Problem 10.1

Q Using a **Born–Haber cycle**, and the following thermodynamic data, calculate the enthalpy of formation ($\Delta_f H°$) for caesium neonide (Cs^+Ne^-), and thus comment on whether it might be a stable compound.

Compare Worked Problems 3.2 and 4.1.

$$\Delta H \text{ (kJ mol}^{-1}\text{)}$$

	ΔH (kJ mol^{-1})	
$Cs_{(s)} \rightarrow Cs_{(g)}$	+79	enthalpy of sublimation
$Cs_{(g)} \rightarrow Cs^+_{(g)}$	+378	first ionization energy
$Ne_{(g)} \rightarrow Ne^-_{(g)}$	+31	electron affinity
$Cs^+_{(g)} + Ne^-_{(g)} \rightarrow Cs^+Ne^-_{(s)}$	–567	lattice enthalpy (Section 3.4); calculated

First ionization energy of Cs: Figure 3.1.

A Construct a Born–Haber cycle such that the unknown parameter ($\Delta_f H°$) completes the closed cycle:

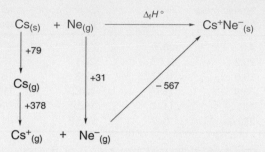

Hence $\Delta_f H° = 79 + 378 + 31 - 567 = -79$ kJ mol^{-1}. With a negative enthalpy of formation, though small, the compound might be stable.

10.2 The Elements

10.2.1 Occurrence

Uranium isotopes undergo radioactive decay to give α-particles (^{4}He nuclei), *e.g.*
^{238}U $\rightarrow \rightarrow \rightarrow$ ^{206}Pb + 8α + 6β
(β = electron)

The noble gases are used as low-temperature refrigerants, to provide an inert atmosphere, and liquid xenon has found applications as an extremely unreactive solvent.

All of the elements exist as highly unreactive monatomic gases, with very low boiling points (Table 10.1). Argon is the most abundant, comprising 0.93% of the atmosphere (see Figure 7.2). Helium, neon, krypton and xenon are trace constituents of air, and with the exception of helium are obtained by fractional distillation of liquid air. All isotopes of radon are radioactive, and are formed naturally from the decay of heavy radioisotopes such as those of radium and uranium.[3]

10.2.2 Noble Gas Clathrates

In the 1920s and 1930s there was considerable interest in materials formed between noble gases (particularly Ar, Kr and Xe) and strongly hydrogen-bonding compounds such as water or polyphenols. These have subsequently been shown to be not chemical compounds of the noble gases, but clathrate compounds (also known as inclusion compounds), where the gas atoms occupy voids in the hydrogen-bonded lattice of the host compound when the host is crystallized under a noble gas atmosphere.

Xenon forms a clathrate with water of approximate composition $Xe(H_2O)_n$ ($n = 5$ or 6), which has a melting point of 24 °C. Figure 10.1 shows a schematic diagram of ice (which normally crystallizes with a diamond-like structure) with an encapsulated noble gas atom. On melting such substances, the noble gas escapes.

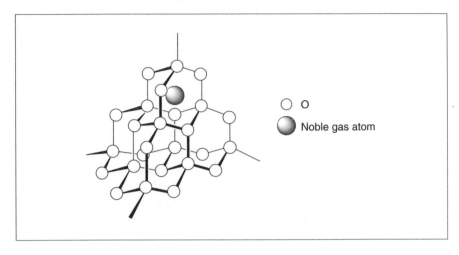

○ O

⬤ Noble gas atom

Figure 10.1 Schematic diagram of ice containing an encapsulated noble gas atom in a void in the lattice. Hydrogen atoms of water molecules are omitted for clarity

10.3 Noble Gas Halides

These are the most important compounds of the noble gases; fluorine is the most electronegative element and is able to stabilize the highly oxidizing noble gas compounds. The vast majority of compounds are fluorides of xenon (and to a far lesser extent of krypton), though some chemistry of radon is known, hampered by its radioactivity and availability in only small quantities.

Fluorides are the most important compounds formed by the noble gases.

10.3.1 Krypton Difluoride

For krypton, the only known halide is KrF_2, which is synthesized by passing an electrical discharge through a fluorine–krypton mixture. KrF_2 is a powerful fluorinating agent, and is more reactive than XeF_2. It will oxidize Xe to XeF_6 and metallic gold to AuF_6^- (equations 10.1 and 10.2).

$$3KrF_2 + Xe \rightarrow XeF_6 + 3Kr \qquad (10.1)$$

$$7KrF_2 + 2Au \rightarrow 2KrF^+AuF_6^- + 5Kr \qquad (10.2)$$

10.3.2 Xenon Fluorides

A xenon fluoride, 'XePtF$_6$', was the first noble gas compound to be synthesized by reaction of Xe with the strong oxidant PtF_6, and later shown to be probably a mixture of $[XeF]^+[PtF_6]^-$ and $[XeF]^+[Pt_2F_{11}]^-$.[4] This heralded the start of intensive investigations into noble gas chemistry, which has continued to the present day.

Bartlett[5] investigated the reaction of xenon and PtF_6 since the latter was able to oxidize dioxygen (O_2) to the oxygenyl cation (O_2^+), and the first ionization energies of O_2 (914 kJ mol^{-1}) and Xe (1177 kJ mol^{-1}) are quite similar. In comparison, the first ionization energies for the other noble gases are much higher (see Table 10.1).

Conditions for the synthesis of XeF_4 and XeF_6:
XeF_4: Xe + $5F_2$, 400 °C, 5 atmospheres pressure.
XeF_6: Xe + $20F_2$, 300 °C, 50 atmospheres pressure.

Xenon forms three neutral fluorides, XeF_2, XeF_4 and XeF_6, by the reaction of fluorine and xenon under different conditions. The simplest synthesis of XeF_2 is to expose a fluorine–xenon mixture (contained in a dry glass bulb) to sunlight; colourless crystals of XeF_2 are deposited on the walls of the flask. The sunlight causes dissociation of the relatively weak F–F bond in elemental fluorine to form fluorine atoms, which then react with the xenon. With larger mole ratios of fluorine to xenon, higher temperatures and higher pressures, XeF_4 and XeF_6 can be synthesized.

XeF_2 and XeF_4 have the expected shapes predicted by VSEPR (see Section 1.4), as shown in Figure 10.2. XeF_6 is a seven electron-pair molecule, and may theoretically have an octahedral structure (where the lone pair is in a spherically inactive s-orbital) or a distorted octahedral structure (where the lone pair is active). In fact, XeF_6 is fluxional in the gas phase, interchanging between structures where the lone pair points through the centre of an F_3 triangle (one face) of a distorted octahedral XeF_6 molecule (Figure 10.2).

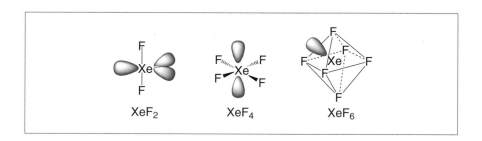

Figure 10.2 The shapes of xenon fluorides, showing the positions of the lone pairs

XeF_2 XeF_4 XeF_6

All of the xenon fluorides are powerful fluorinating agents, able to oxidize a wide range of compounds. In many cases XeF_2 is a very *selective* oxidant, able to oxidize the central heteroatom of a main group compound (such as As, P, *etc.*) but not organic substituents (*e.g.* methyl, phenyl) bonded to it, as illustrated in equations 10.3 and 10.4.

$$Me_3As + XeF_2 \rightarrow Me_3AsF_2 + Xe \qquad (10.3)$$

$$Ph_2PH + XeF_2 \rightarrow Ph_2PHF_2 + Xe \qquad (10.4)$$

XeF_2 also oxidizes water to oxygen (equation 10.5), while XeF_4 can oxidize platinum metal to PtF_4 (equation 10.6).

$$2H_2O + 2XeF_2 \rightarrow O_2 + 4HF + 2Xe \qquad (10.5)$$

$$XeF_4 + Pt \rightarrow PtF_4 + Xe \qquad (10.6)$$

10.4 Reactions of Noble Gas Fluorides with Fluoride Ion Acceptors and Donors

There is an extensive chemistry where noble gas fluorides react with strong fluoride ion acceptors, such as the pentafluorides MF_5 of As, Sb, Bi, Ta, Ru or Pt. XeF_2 forms the greatest number of compounds by this type of reaction, followed by XeF_6 and then XeF_4. KrF_2 forms many similar compounds to XeF_2. In some cases the fluoride anion is completely transferred to the Lewis acid MF_5, leaving cationic xenon species (for example, XeF_2 gives the XeF^+ cation which can be solvated, *e.g.* by a nitrile RCN; see Section 9.6), but in most cases the fluoride is only partly transferred, resulting in compounds with Xe–F–M fluoride bridges (Figure 10.3a). The XeF^+ cation is also able to lose its fluoride ion on reaction with reducing agents, or with a mixture of HF and SbF_5 under a xenon atmosphere, to give the green $[Xe_2]^+$ cation.[6]

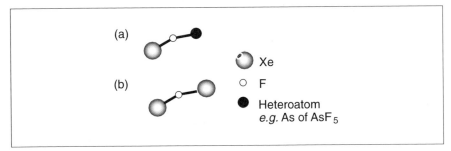

(a)

(b)

Xe

F

Heteroatom
e.g. As of AsF_5

Figure 10.3 Types of fluoride bridges involving xenon

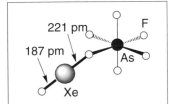

221 pm

187 pm

F

As

Xe

Figure 10.4 Structure of $[XeF]^+[AsF_6]^-$

Different types of cations and anions tend to be obtained depending on the stoichiometry of noble gas fluoride to MF_5. As an example of a compound formed from a 1:1 ratio, solid XeF_2 reacts with liquid AsF_5 (equation 10.7) to give a salt with fluoride bridging between the $[XeF]^+$ cation and $[AsF_6]^-$ anion (Figure 10.4).

$$XeF_{2(s)} + AsF_{5(l)} \rightarrow [XeF]^+[AsF_6]^-_{(s)} \qquad (10.7)$$

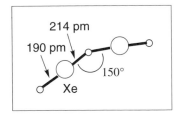

214 pm

190 pm

150°

Xe

Figure 10.5 Structure of the $[Xe_2F_3]^+$ cation

When a 2:1 ratio of noble gas compound to MF_5 is used, the adducts tend to contain cations with Xe–F–Xe bridges (Figure 10.3b), such as the $[Xe_2F_3]^+$ cation which occurs in $[Xe_2F_3]^+[AsF_6]^-$. The $[Xe_2F_3]^+$ cation (Figure 10.5) can alternatively be viewed as an XeF^+ cation solvated by XeF_2. For compounds containing a 1:2 molar ratio of noble gas fluoride to MF_5, they tend to contain $[M_2F_{11}]^-$ anions (*e.g.* $Sb_2F_{11}^-$, Figure 10.6) which are linked to the noble gas cation (*e.g.* XeF^+) by a fluoride bridge. Noble gas fluorides are able to act as fluoride acceptors themselves, and will add one or two fluoride anions. Examples include the addition of F^- to XeF_4 to give the pentagonal planar $[XeF_5]^-$ anion, or to XeF_6 to give either $[XeF_7]^-$ or $[XeF_8]^{2-}$, depending on the amount of fluoride added.

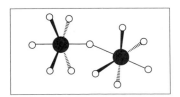

Figure 10.6 Structure of the $[Sb_2F_{11}]^-$ anion

> ### Worked Problem 10.2
>
> **Q** (a) How would the compound $[XeF_3][AsF_6]$ be prepared?
> (b) What is the nature of the compound of stoichiometry $KrF_2.2SbF_5$?
>
> **A** (a) To prepare this compound, fluoride transfer from Xe to As has occurred in the reaction between XeF_4 and AsF_5.
> (b) Here, fluoride has been transferred from KrF_2 to SbF_5. Two SbF_5 molecules share a fluoride anion, giving $Sb_2F_{11}^-$; hence the compound is likely to be $[KrF]^+[Sb_2F_{11}]^-$.

10.5 Xenon–Oxygen Compounds

These compounds generally result from hydrolysis of the fluorides XeF_4 and XeF_6. As described in equation 10.5, hydrolysis of XeF_2 results in oxidation of water to oxygen, and formation of xenon gas. However, when XeF_4 is hydrolysed, some xenon gas is liberated and some XeF_4 is converted to xenon trioxide (XeO_3) which remains in solution. In this reaction the xenon(IV) (in XeF_4) is undergoing disproportionation to $Xe(0)$ (in xenon gas) and $Xe(VI)$ (in XeO_3) (equation 10.8).

$$6XeF_4 + 12H_2O \rightarrow 4Xe + 2XeO_3 + 24HF + 3O_2 \qquad (10.8)$$

XeO_3 is also obtained by the hydrolysis of XeF_6 (equation 10.9).

$$XeF_6 + 3H_2O \rightarrow XeO_3 + 6HF \qquad (10.9)$$

Shape of XeO_3

XeO_3 is a highly explosive white solid, soluble in water in which it appears not to be ionized. In strong alkaline solution, however, XeO_3 behaves as a weak acid, giving the xenate(VI) anion, $[XeO_3(OH)]^-$, by addition of OH^- (equation 10.10).

$$XeO_3 + OH^- \rightleftharpoons [XeO_3(OH)]^- \qquad (10.10)$$

Shape of XeO_6^{4-}

The xenate(VI) anion is unstable in aqueous solution, and undergoes a disproportionation reaction to give xenon gas and the xenate(VIII) anion, XeO_6^4 (equation 10.11), which contains xenon in the +8 oxidation state.

$$2XeO_3(OH)^- + 2OH^- \rightarrow XeO_6^{4-} + Xe + O_2 + 2H_2O \quad (10.11)$$

Acidification of xenate(VIII) solutions gives xenon tetroxide, XeO_4, again with xenon in the +8 oxidation state. XeO_4 is highly unstable, decomposing rapidly to xenon gas plus O_2, often explosively.

Oxyfluorides are also well-known species. The xenon(VI) compound $XeOF_4$ can be prepared by partial hydrolysis of XeF_6 (equation 10.12), but this route is less favoured than the reaction of XeF_6 with sodium nitrate (equation 10.13) or the reaction of XeF_6 and POF_3 (equation 10.14) since they do not generate the explosive XeO_3.

$$XeF_6 + H_2O \rightarrow XeOF_4 + 2HF \qquad (10.12)$$

$$XeF_6 + NaNO_3 \rightarrow XeOF_4 + NaF + FNO_2 \qquad (10.13)$$

$$XeF_6 + POF_3 \rightarrow XeOF_4 + PF_5 \qquad (10.14)$$

The reaction of XeF_6 with Na_4XeO_6 gives the oxyfluorides $XeOF_4$ and XeO_3F_2; XeO_2F_2 is also known.

Worked Problem 10.3

Q Similar to xenon fluorides (Section 10.4), oxyfluorides are also able to form oxyfluoro cations and anions by reaction with fluoride ion acceptors and donors, respectively. Suggest syntheses for the following ions, and predict their shapes using VSEPR: (a) $[XeO_2F]^+$; (b) $[XeOF_3]^+$.

A (a) $[XeO_2F]^+$ can be formed from XeO_2F_2 by removal of F^-, using a suitable acceptor such as AsF_5:

$$XeO_2F_2 + AsF_5 \rightarrow [XeO_2F]^+[AsF_6]^-$$

The Xe atom has 8 valence electrons, plus 3 from σ-bonds to O and F, minus two electrons for Xe=O π-bonds, minus one electron for the positive charge, giving a total of 8 electrons, *i.e.* 4 pairs. The electron pairs adopt a tetrahedral arrangement, and the molecule has a trigonal pyramidal shape, with one lone pair.

(b) $[XeOF_3]^+$ can be formed from $XeOF_4$ by fluoride removal:

$$XeOF_4 + SbF_5 \rightarrow [XeOF_3]^+[SbF_6]^-$$

The Xe atom has 8 valence electrons, plus 4 from σ-bonds to O and F, minus one for the Xe=O π-bond, minus one for the charge, giving a total of 10 electrons. The arrangement of electron pairs is trigonal bipyramidal with one lone pair and an Xe=O double bond; both will be in equatorial positions, and the F–Xe–F angle will be <180° (refer to Section 1.4.5 for the shapes of trigonal bipyramidal species).

Other xenon–oxygen compounds can be formed from XeF_2, XeF_4 and XeF_6 by substitution of one or more fluorides by reaction of strong oxyacids (XOH) such as CF_3SO_3H, $MeSO_3H$, CF_3CO_2H, FSO_3H or TeF_5OH, with elimination of HF (equation 10.15).

$$F–Xe–F + XOH \rightarrow F–Xe–OX \ (+ \ HF) \rightarrow XO–Xe–OX \ (+ \ HF)$$
$$(10.15)$$

Alternatively, fluoride exchange yields analogous products, *e.g.* equation 10.16.

$$XeF_6 + 2B(OTeF_5)_3 \rightarrow Xe(OTeF_5)_6 + 2BF_3 \qquad (10.16)$$

The compounds have similar shapes to the fluorides, *e.g.* XO–Xe–OX and XeF_2 are both linear.

10.6 Xenon and Krypton Compounds with Bonds to Elements other than O and F

While the initial studies on xenon chemistry revealed the existence of compounds with small, highly electron-withdrawing atoms (namely oxygen and fluorine) bonded to xenon, in recent years there has been a surge in activity on the synthesis of compounds with bonds between xenon and elements of lower electronegativity. In order to stabilize such compounds, electron-withdrawing groups, typically with fluorine substitution, are often necessary. Thus, the highly electron-withdrawing amide $[(FSO_2)_2N]^-$ forms the xenon compound $FXeN(SO_2F)_2$, which has a linear F–Xe–N group. The XeF^+ and KrF^+ cations have also been isolated as species solvated by nitriles such as $[(C_2F_5CN)XeF]^+$ and $[(HCN)KrF]^+$.[7]

Xenon–carbon bonded compounds include pentafluorophenyl species such as $[C_6F_5Xe]^+$, obtained by transfer of pentafluorophenyl groups from boron to xenon (equation 10.17).

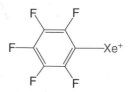

The pentafluorophenylxenon cation $[C_6F_5Xe]^+$

In MeCN solution, the $[C_6F_5Xe]^+$ cation is solvated to give $[C_6F_5Xe(NCMe)]^+$, which has the expected linear N–C–Xe geometry.

$$XeF_2 + B(C_6F_5)_3 \rightarrow [C_6F_5Xe]^+[C_6F_5BF_3]^- + [C_6F_5Xe]^+[(C_6F_5)_2BF_2]^-$$
$$(10.17)$$

Alkynyl species of the type $[R–C{\equiv}C–Xe]^+$, where R can be a range of non-fluorinated groups such as ethyl or trimethylsilyl ($SiMe_3$) have been made. Thus, there appears to be a rather extensive organic chemistry of xenon which awaits discovery.

In contrast, few compounds containing bonds from xenon to other elements have been synthesized. Reactions of the $[C_6F_5Xe]^+$ cation with Cl^- gives C_6F_5XeCl, and with Me_3SiCl gives $[(C_6F_5Xe)_2Cl]^+$.[8]

Summary of Key Points

1. *Xenon* is the only element to show an *extensive chemistry*, which begins with the binary fluorides XeF_2, XeF_4 and XeF_6, and which can be classified into:
   - Reactions as a *fluorinating agent*
   - *Hydrolysis* reactions
   - *Exchange* reactions of fluorine for another anion
   - Reactions with *fluoride donors* and *acceptors*.

2. *Compounds* containing bonds from *Xe to F, O, C, N and Cl* are all well characterized.

3. *Few compounds of radon or krypton are known;* all are in the +2 oxidation state, and those of krypton are more reactive than their xenon analogues.

Problems

10.1. Determine the oxidation state of xenon in the following species:
(a) $[XeO_2F_3]^-$; (b) $[F(XeOF_4)_3]^-$; (c) $[Bu^t–C≡C–Xe]^+$; (d) $O_2Xe(OTeF_5)_2$.

10.2. Identify the Group 18 element X in each of the following:
(a) X is an inert gas used in coloured lighting.
(b) X is the most abundant of the noble gas elements.
(c) X forms compounds only in the +2 oxidation state.
(d) X is a radioactive gas.

10.3. Predict the products of the following reactions where XeF_2 is present as a fluorinating agent in excess, and write balanced equations:
(a) $C_6F_5I + XeF_2 \rightarrow$
(b) $Ph_2S + XeF_2 \rightarrow$ a sulfur(IV) compound
(c) $S_8 + XeF_2 \rightarrow$

10.4. The following compounds have recently been synthesized. Suggest a possible method to achieve the synthesis.
(a) $Kr(OTeF_5)_2$ [the first example of a compound with a Kr–O bond].
(b) $[XeOF_5]^-$ [an anion with a pentagonal-pyramidal structure].

10.5. An improved method for the synthesis of pure XeF_4 from Xe uses the powerful oxidant O_2F_2. Write a balanced equation for the reaction, assuming the by-product is dioxygen.

10.6. Using the following data, together with those in Worked Problem 10.1, calculate the enthalpies of formation ($\Delta_f H°$) of caesium xenonide (Cs^+Xe^-) and potassium neonide (K^+Ne^-), and comment on the values:

	ΔH (kJ mol^{-1})
$K_{(s)} \rightarrow K_{(g)}$	+90
$K_{(g)} \rightarrow K^+_{(g)}$	+421
$Xe_{(g)} \rightarrow Xe^-_{(g)}$	+43
$Cs^+_{(g)} + Xe^-_{(g)} \rightarrow Cs^+Xe^-_{(s)}$	−457
$K^+_{(g)} + Ne^-_{(g)} \rightarrow K^+Ne^-_{(s)}$	−613

10.7. Using VSEPR (see Section 1.4), explain why:
(a) the $[XeF_5]^-$ ion is planar;
(b) the $[(C_6F_5Xe)_2Cl]^+$ ion (Section 10.6) has a bent Xe–Cl–Xe group.

References

1. J. H. Holloway, *Chem. Br.*, 1987, 658; P. Laszlo and G. J. Schrobilgen, *Angew. Chem., Int. Ed. Engl.*, 1988, **27**, 479.
2. G. H. Purser, *J. Chem. Educ.*, 1988, **65**, 119.
3. J. D. Lee and T. E. Edmonds, *Educ. Chem.*, 1991, 152.
4. F. O. Sladky, P. A. Bulliner and N. Bartlett, *J. Chem. Soc (A)*, 1969, 2179.
5. N. Bartlett, *Proc. Chem. Soc.*, 1962, 218.
6. T. Drews and K. Seppelt, *Angew. Chem., Int. Ed. Engl.*, 1997, **36**, 273.
7. W. Koch, *J. Chem. Soc., Chem. Commun.*, 1989, 215.
8. H.-J. Frohn, T. Schroer and G. Henkel, *Angew. Chem., Int. Ed. Engl.*, 1999, **38**, 2554.

Further Reading

Recent advances in noble gas chemistry, J. H. Holloway and E. G. Hope, *Adv. Inorg. Chem.*, 1999, **46**, 51.
Twenty-five years of noble gas chemistry, J. H. Holloway, *Chem. Br.*, 1987, 658.
A noble cause, G. M. R. Grant, *Chem. Br.*, 1994, 388.

11

The Group 12 Elements: Zinc, Cadmium and Mercury

Aims

By the end of this chapter you should understand:

- The similarities and differences between these elements and traditional main group elements
- The dominance of the +2 oxidation state in the chemistry of these elements
- The strong affinity of mercury for sulfur ligands

First and second ionization energies (kJ mol^{-1}):

Element	First	Second
Zn	+913	+1740
Cd	+874	+1638
Hg	+1013	+1816
Be	+906	+1763

11.1 Introduction

Zinc, cadmium and mercury, while formally part of the d-block series of elements, have properties much more in keeping with main group elements than transition metals, and are therefore often considered in discussions of the former. All three elements have the $d^{10}s^2$ electronic configuration, and the group oxidation state is +2, formed by loss of the two s-electrons, though it has been speculated that Hg(IV) may be obtainable as HgF_4.[1] In many cases, zinc and cadmium compounds are similar to the analogous magnesium compounds, and often have the same structures. Mercury compounds have a greater degree of covalent character, often with low coordination numbers. The +1 oxidation state is well established for mercury, but is very unstable for zinc and cadmium.

Zinc shows several similarities to beryllium: both metals have similar first and second ionization energies, though the Zn^{2+} ion is larger than Be^{2+}, often resulting in differences in coordination numbers in complexes (zinc can be four, five or six coordinate, but beryllium has a maximum coordination number of four). Thus, for example, both metals react with acids and alkalis, and both form a basic ethanoate (acetate) of the type $M_4O(O_2CMe)_6$ (see Section 4.3).

11.2 The Elements

Because the ns electrons in Zn, Cd and Hg are tightly bound, they are relatively unavailable for metallic bonding and so the metals are volatile, with low melting and boiling points. Mercury is a unique metal, being a liquid at room temperature and forming a monoatomic gas.

Metal	Mp (°C)	Bp (°C)
Zn	420	907
Cd	320	767
Hg	–39	357

Zinc, cadmium and mercury are chalcophilic, that is they show a strong affinity for the chalcogens sulfur, selenium and tellurium. They occur naturally as sulfide minerals, in a concentrated form, from which they are easily extracted. The main ore of zinc is zinc blende (*sphalerite*, ZnS), which often contains cadmium. Mercury occurs as *cinnabar* (HgS). The sulfide ores are roasted in air, whereupon zinc and cadmium sulfides are converted to the oxide (equation 11.1). HgS is unstable above 400 °C and decomposes to the metal (equation 11.2). The ZnO (and CdO) are reduced to the metals using carbon (equation 11.3).

$$2ZnS + 3O_2 \rightarrow 2ZnO + 2SO_2 \tag{11.1}$$

$$HgS + O_2 \rightarrow Hg + SO_2 \tag{11.2}$$

$$ZnO + C \rightarrow Zn + CO \tag{11.3}$$

11.3 Chemistry of the Elements

The reactivity of the metals decreases going down the group; although zinc and cadmium have high first ionization energies, their redox potentials are quite large and negative (Table 11.1) owing to the high solvation energy that drives the reaction, so they readily dissolve in non-oxidizing acids (equation 11.4). In contrast, mercury will only dissolve in oxidizing acids, such as nitric acid.

$$Zn_{(s)} + 2H^+_{(aq)} \rightarrow Zn^{2+}_{(aq)} + H_{2(g)} \tag{11.4}$$

Table 11.1 Standard redox potentials for zinc, cadmium and mercury at 25 °C

Reaction	E° (V)
$Zn^{2+} + 2e^- \rightleftharpoons Zn$	–0.76
$Cd^{2+} + 2e^- \rightleftharpoons Cd$	–0.40
$Hg^{2+} + 2e^- \rightleftharpoons Hg$	+0.85

11.4 Halides

All combinations of halide and metal are known for the group. The fluorides are ionic with high melting points. ZnF_2 crystallizes with the rutile structure (see Figure 6.9), while CdF_2 and HgF_2 have the fluorite structure (see Figure 4.2). ZnF_2 and CdF_2 resemble MgF_2 in being poorly soluble in water. The chlorides, bromides and iodides of zinc and cadmium are largely ionic, but with increasing covalent character for the combination of cadmium with heavier halides. In contrast, mercury(II) halides (HgX_2) are covalent solids, which are only slightly soluble in water and are only slightly dissociated into Hg^{2+} and X^- ions.

11.5 Chalcogenides and Related Compounds

From zinc to cadmium to mercury, there is an increasing tendency to form stable compounds with the chalcogens S, Se and Te. The chalcogenides can be formed either by direct combination, *e.g.* equation 11.5, or by reaction of aqueous M^{2+} ions with H_2S, H_2Se or H_2Te (or a salt thereof, *e.g.* Na_2S) (equation 11.6).

$$Hg + S \rightarrow HgS \qquad (11.5)$$

$$Cd(NO_3)_2 + H_2S \rightarrow CdS + 2HNO_3 \qquad (11.6)$$

ZnS crystallizes in two forms: the low-temperature zinc blende (sphalerite) form and the higher-temperature wurtzite form. The sphalerite structure (Figure 11.1) is related to that of diamond (see Figure 6.3) with a cubic close-packed lattice of sulfur atoms, and zinc atoms in half of the tetrahedral holes (there are two holes per close-packed atom). The wurtzite structure (Figure 11.2) is based on a hexagonal close-packed lattice of sulfur atoms, again with zinc atoms in half the tetrahedral holes.

Mercury (and, to a lesser extent, cadmium and zinc) has a very strong affinity for thiolate ligands, RS^-; indeed, thiols (RSH) have been traditionally known as mercaptans for this very reason. Thiolate complexes of zinc, cadmium and mercury are very important in biological systems; for example, in enzymes zinc is often bonded to the S atom of the amino acid cysteine.

ZnS is an important phosphor used in TV screens.

$$HS-CH_2-CH \begin{array}{c} NH_2 \\ \\ CO_2H \end{array}$$

Cysteine

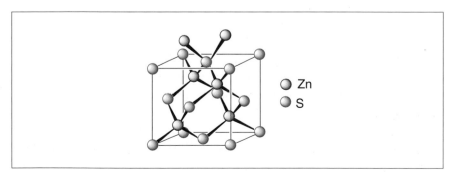

Figure 11.1 The structure of the *sphalerite* form of ZnS

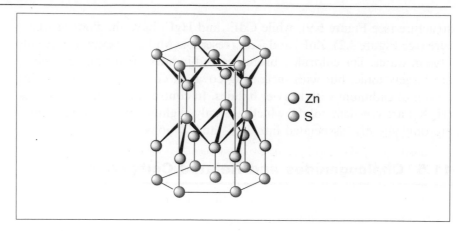

Figure 11.2 The structure of the *wurtzite* form of ZnS

11.6 Oxygen Compounds

Zinc has a relatively high affinity for oxygen, whereas mercury does not.

The metals react directly with oxygen on heating; however, above 400 °C, HgO decomposes back to the metal plus oxygen. ZnO adopts the wurtzite (ZnS) structure (Figure 11.2) which has tetrahedral coordination of the zinc ions, whereas in CdO the larger Cd^{2+} ion is better accommodated by the sodium chloride structure (see Figure 3.4), which has six-coordinate Cd ions.

ZnO is amphoteric, dissolving in both acids and bases. With excess base, the hydrated zincate anions $[Zn(OH)_x(H_2O)_y]^{(x-2)-}$ are formed, and solid salts such as $Na[Zn(OH)_3]$ and $Na_2[Zn(OH)_4]$ can be crystallized. In contrast, CdO is not amphoteric. However, analogous cadmate anions may be formed in *small* amounts, since $Cd(OH)_2$ dissolves in hot and very concentrated KOH solution, though no cadmate salts can be isolated. Addition of hydroxide to solutions of M^{2+} ions precipitates $Zn(OH)_2$ and $Cd(OH)_2$; however, $Hg(OH)_2$ does not exist and instead yellow HgO is formed (equations 11.7 and 11.8). Zinc and cadmium hydroxides also dissolve in aqueous ammonia solutions by the formation of ammonia complexes such as $[Zn(NH_3)_4(H_2O)_2]^{2+}$.

$$Zn^{2+} + 2OH^- \rightarrow Zn(OH)_2 \qquad (11.7)$$

$$Hg^{2+} + 2OH^- \rightarrow HgO + H_2O \qquad (11.8)$$

A wide range of salts of oxyacids are known for all three metals, including nitrates, sulfates, chlorates(VII), *etc.* Many are easily synthesized by reaction of the oxide with the acid, followed by crystallization, *e.g.* equation 11.9.

$$ZnO + 2HClO_4 + 5H_2O \rightarrow Zn(ClO_4)_2.6H_2O \qquad (11.9)$$

Zinc and cadmium carbonates are rather unstable to heat as a result of the polarizing effect of the small Zn^{2+} and Cd^{2+} ions, causing decomposition to the oxide (equation 11.10). In this way they resemble magnesium carbonate (see Section 4.3).

$$CdCO_3 \rightarrow CdO + CO_2 \qquad (11.10)$$

Worked Problem 11.1

Q How could you distinguish samples of $Zn(NO_3)_2$ and $MgSO_4$?

A Addition of sodium hydroxide solution would produce white precipitates [of $Mg(OH)_2$ or $Zn(OH)_2$], but the $Zn(OH)_2$ would dissolve in excess OH^- to give a clear, colourless solution. Alternatively, adddition of aqueous $Ba(NO_3)_2$ would produce a precipitate of $BaSO_4$ with $MgSO_4$.

11.7 Formation of Coordination Complexes

The formation of coordination complexes by the zinc group elements is a trait more allied with the transition metals than with main group elements, illustrating the intermediate character of the group. Tetrahedral four-coordination is common among the complexes, which are formed from a wide range of donor ligands. Examples include cationic complexes such as $[Zn(NH_3)_4]^{2+}$, neutral complexes such as $[HgCl_2(PPh_3)_2]$ and $[Zn(pyridine)_2Cl_2]$ and anionic complexes such as $[Zn(CN)_4]^{2-}$ and $[CdI_4]^{2-}$. Complexes of cadmium are often more stable when in six coordination. Many hydrated salts of strong acids [*e.g.* $Zn(ClO_4)_2$, $Cd(BF_4)_2$] contain the $[Zn(H_2O)_6]^{2+}$ or $[Cd(H_2O)_6]^{2+}$ ions.

Mercury has a strong tendency to adopt lower coordination numbers than cadmium or zinc; it often displays linear two-coordination such as in the organometallic derivatives R_2Hg and $RHgX$ (X = halide), though four-coordination is quite common, *e.g.* the colourless $[HgI_4]^{2-}$ formed when red HgI_2 dissolves in KI solution (equation 11.11).

$$HgI_2 + 2I^- \rightarrow [HgI_4]^{2-} \qquad (11.11)$$

See also Section 11.6.

Pyridine

> ### Worked Problem 11.2
>
> **Q** Can VSEPR (see Section 1.4) be used to predict the shapes of complexes of the zinc group dipositive cations? Give an example.
> **A** The M^{2+} cations of the zinc group metals have a filled d-shell (the cations have lost the s-electrons), so VSEPR can be used to predict shapes. For example, $[Zn(NH_3)_4]^{2+}$: Zn has no valence electrons, 4 dative bonds from NH_3 ligands contributes 4 electron pairs, which are tetrahedrally arranged. Similarly, in PhHgCl, there are 2 valence pairs of electrons, so the molecule is linear.

11.8 Low-valent Compounds

See Section 1.5 for discussion on paramagnetism and diamagnetism.

The Cd_2^{2+} and Zn_2^{2+} species are known, but unstable; they are formed only in anhydrous conditions and readily disproportionate in water to give the metal and Zn^{2+} or Cd^{2+}. In contrast, the Hg_2^{2+} ion is the most stable, and the best known example of a monovalent species formed by the group. The Hg_2^{2+} ion is *diamagnetic* (with no unpaired electrons); an M^+ ion would be paramagnetic with one unpaired electron, so dimerization occurs through Hg–Hg bond formation, giving the Hg_2^{2+} ion. There is an equilibrium between Hg_2^{2+} and Hg^{2+} (equation 11.12), which has an equilibrium constant $[Hg_2^{2+}]/[Hg^{2+}]$ of about 170.

$$Hg^{2+} + Hg \rightleftharpoons Hg_2^{2+} \tag{11.12}$$

Thus, in an aqueous solution of a mercury(I) salt there will be >0.5% of Hg^{2+} present in solution. If Hg^{2+} ions are removed by complexation with ligands which form stable complexes or insoluble compounds with it (such as cyanide), then the disproportionation goes to completion, *e.g.* equation 11.13.

$$Hg_2^{2+}{}_{(aq)} + 2CN^-{}_{(aq)} \rightarrow Hg_{(l)} + Hg(CN)_{2(s)} \tag{11.13}$$

> ### Worked Problem 11.3
>
> **Q** Using the following thermodynamic data, show that Hg_2Cl_2 is unstable with respect to disproportionation to $HgCl_2$ and Hg:
> $\Delta_f H°(Hg_2Cl_2)$ –265 kJ mol^{-1}, $\Delta_f H°(HgCl_2)$ –224 kJ mol^{-1}.
>
> **A** For the equation $Hg_2Cl_2 \rightarrow HgCl_2 + Hg$, $\Delta H = \Delta_f H°(HgCl_2) + \Delta_f H°(Hg) - \Delta_f H°(Hg_2Cl_2) = -224 - (-265) - (0) = -41$ kJ mol^{-1}, so the disproportionation reaction is favoured.

Worked Problem 11.4

Q Which of the following properties of transition elements are displayed by zinc: (a) variable oxidation states; (b) formation of coloured ions; (c) formation of complex ions; (d) formation of paramagnetic ions.

A The only similarity among the list between zinc and transition metals is the formation of complex ions (Section 11.8). Zinc otherwise forms colourless Zn^{2+} ions which have a filled d-shell and are diamagnetic.

Summary of Key Points

1. *Many similarities* are shown between the zinc group and main group metals, specifically the heavier p-block metals. Zinc shows some similarities with beryllium.

2. *Mercury is a unique element* in a number of respects: the liquid state of the element, the tendency for low coordination numbers and the very high affinity for sulfur compounds.

Problems

11.1. What is the oxidation state of mercury in the following compounds or ions: (a) HgSe; (b) HgI_4^{2-}; (c) Hg_2^{2+}; (d) $[HgCl_2(PPh_3)_2]$.

11.2. Identify the element X in each of the following;
(a) The oxide XO decomposes on heating.
(b) The organometallic compound Me_2X is stable to water.
(c) The oxide XO dissolves in strong NaOH solution.

11.3. Zinc shows amphoteric-like behaviour, and the metal dissolves in both acids and alkalis, giving hydrogen gas, and in alkali gives the $[Zn(OH)_4]^{2-}$ ion. The behaviour of zinc with acids and bases in liquid ammonia (see Section 7.3.1) is analogous to that in water. Write balanced equations for reactions of metallic zinc with:
(a) aqueous HCl; (b) excess aqueous sodium hydroxide; (c) NH_4Cl in liquid ammonia; (d) excess $NaNH_2$ in liquid ammonia.

11.4. Zinc, cadmium and mercury have a strong affinity for sulfur and selenium. Which other p-block elements occur naturally as sulfides?

11.5. Predict the outcome of the following reactions:
(a) $Hg_2^{2+} + Na_2S \rightarrow$
(b) $CdCl_{2(aq)} + Na_2CO_{3(aq)} \rightarrow$

Reference

1. M. Kaupp, M. Dolg, H. Stoll and H. G. von Schnering, *Inorg. Chem.*, 1994, **33**, 2122.

12

Selected Polymeric Main Group Compounds

Aims

By the end of this chapter you should understand:

- The diversity of polymers formed by main group elements
- Structural features of silicates and polyphosphates
- The synthesis of some important synthetic main group polymers

12.1 Introduction

Modern society utilizes a wide variety of polymers; these occur in many natural and synthetic forms, with different properties. Historically, synthetic polymers have been largely organic-based, and the development of synthetic inorganic polymers has lagged far behind. While some polymeric inorganic materials, *e.g.* the silicates, have existed for billions of years, it is only recently that there has been an upsurge in activity in the synthesis of novel inorganic polymers which possess interesting properties, some of which may be complementary to those of organic polymers.

The purpose of this chapter is to discuss some of the more important types of polymers and both natural silicates and a selection of synthetic polymers are described. A polymer is here considered as a material built up from essentially covalent bonds, so that ionic substances and metals are excluded but silicates and polyphosphates are included.

Other polymeric systems:
sulfur (Chapter 8)
carbon (Section 6.2.1)
boron nitride (Section 5.8.2)
borates (Section 5.7.1)
electron-deficient compounds
(Sections 4.3, 4.7 and 5.6.2)

12.2 Polyphosphates

If acid phosphates, such as NaH_2PO_4 or Na_2HPO_4, containing P–OH groups are heated, a condensation reaction (eliminating water) occurs, with the formation of a P–O–P linkage, as shown schematically in equation 12.1.

$$P–O–H + H–O–P \rightarrow P–O–P + H_2O \qquad (12.1)$$

Cyclic phosphates are also called metaphosphates (compare metasilicates, Section 12.3.3).

In its simplest form, when disodium hydrogenphosphate (Na_2HPO_4) is heated, the **pyrophosphate** $[P_2O_7]^{4-}$ anion (Figure 12.1) is formed. If sodium dihydrogenphosphate (NaH_2PO_4) is heated, since each phosphorus has two P–OH groups then **cyclic phosphates** $[(PO_3)_n]^{n-}$ can be formed, where n is usually 3 or 4, but may be up to 10. The structures of the **trimetaphosphate** $[P_3O_9]^{3-}$ and **tetrametaphosphate** $[P_4O_{12}]^{4-}$ anions are shown in Figure 12.2.

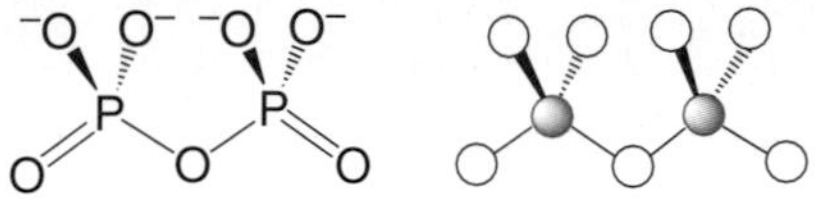

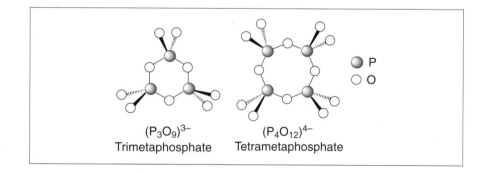

Figure 12.1 The pyrophosphate anion

Figure 12.2 Structures of the cyclic phosphates $[P_3O_9]^{3-}$ and $[P_4O_{12}]^{4-}$

$(P_3O_9)^{3-}$
Trimetaphosphate

$(P_4O_{12})^{4-}$
Tetrametaphosphate

Polyphosphate groups are important in biochemical systems

Linear polyphosphates, containing long chains of $[(PO_3^-)_n]^{n-}$ groups terminated by $–O–PO_3^{2-}$ groups, can also be formed by heating mixtures of HPO_4^{2-} and $H_2PO_4^-$. When trimetaphosphate $[P_3O_9]^{3-}$ is treated with hydroxide ions, one of the P–O–P groups undergoes hydrolysis giving the **triphosphate** $[P_3O_{10}]^{5-}$ ion (Figure 12.3); sodium triphosphate is used in detergents.

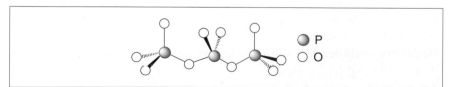

Figure 12.3 Structure of the triphosphate anion, $[P_3O_{10}]^{5-}$

Worked Problem 12.1

Q To form the pentaphosphate ion $[P_5O_{16}]^{7-}$, what ideal molar ratio of Na_2HPO_4 and NaH_2PO_4 should be heated together?

A The pentaphosphate ion will contain two terminal PO_3^{2-} groups, each of which will be derived from a Na_2HPO_4. The remaining three P atoms in the centre of the chain will share two oxygens each; therefore three equivalents of NaH_2PO_4 are needed. Hence the ideal ratio of NaH_2PO_4 to Na_2HPO_4 is 3:2.

An alternative way of representing the phosphate groups in polyphosphates is as a tetrahedral framework unit (Figure 12.4). The tetrametaphosphate $[P_4O_{12}]^{4-}$ ion can therefore be represented as in Figure 12.5. This type of framework representation is widely used for silicate materials (Section 12.3).

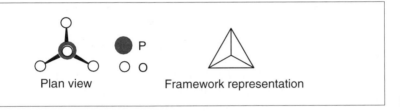

Plan view   ● P   ○ O   Framework representation

Figure 12.4 A PO_4 tetrahedron

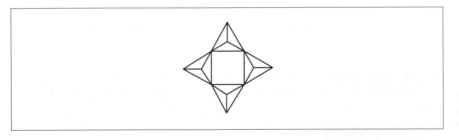

Figure 12.5 Framework representation of the tetrametaphosphate $[P_4O_{12}]^{4-}$ ion

Worked Problem 12.2

Q Draw the framework representation of the triphosphate anion.

A The triphosphate anion is a straight-chain anion comprising three joined PO_4 tetrahedra:

Worked Problem 12.3

Q Explain why there are many polymers derived from phosphoric acid, but none from chloric(VII) acid and few from sulfuric acid.

A Polymer formation can be thought to occur by elimination of H_2O from two E–OH groups, forming an E–O–E linkage (compare equation 12.1). For *chloric(VII) acid* ($HClO_4$), loss of water from two $HClO_4$ molecules forms a Cl–O–Cl linkage, but the resulting Cl_2O_7 molecule (dichlorine heptoxide; see Section 9.5.3) cannot polymerize further. *Sulfuric acid* (H_2SO_4) has two OH groups per molecule, so can theoretically condense to form –S–O–S–O– chains, terminated by OH groups. However, if three H_2SO_4 molecules condense to form a six-membered ring, this is simply one form of sulfur trioxide (see Figure 8.7). In practice, only polysulfate chains containing up to five S atoms are known:

The longest polysulfate chain known

Phosphoric acid (H_3PO_4), however, has *three* OH groups; by condensing two of these the linear and cyclic polyphosphates can be formed, which still retain a P–OH group (or a P–O⁻ group in the salts), giving each polymer an independent identity.

By condensing all three OH groups of phosphoric acid, materials called **ultraphosphates** are formed

12.3 Silicates

12.3.1 Overview

Silicates account for *ca.* 74% of the Earth's crust.

The structural chemistry of silicates is extremely diverse, and these materials are of significant importance as the major rock-forming minerals.

Almost all silicates are based on the SiO_4 *tetrahedron*, which like phosphates is often represented as a framework. There are rare examples of minerals (*e.g.* the *stishovite* form of SiO_2), formed under extreme conditions of temperature and pressure, which have Si in *octahedral* six-coordination. As with polyphosphates, SiO_4 tetrahedra can link together, usually by sharing one to four vertices to form large polysilicate anions. Each terminal oxygen bears a negative charge, but each bridging oxygen is neutral. The negative charge of the silicate anion is neutralized by cations, which, in natural minerals, are often mixtures of cations with similar sizes, for example Mg^{2+} and Fe^{2+}.

Silicates can be classified into various types, according to how the SiO_4 tetrahedra are linked together, and these are discussed in Sections 12.3.2–12.3.5.

Worked Problem 12.4

Q For one ion to substitute another in a silicate mineral, they should ideally have the same charge, and their ionic radii should not differ by >10%. Would you expect the following substitutions to readily occur: (a) Fe^{2+} (78 pm) for Mg^{2+} (72 pm); (b) Li^+ (76 pm) for Na^+ (102 pm).

A (a) The ions have the same charge, and their radii are very close, so substitution would be expected. (b) Although the ions have the same charge, the radius of Na^+ is more than 10% larger than Li^+, so substitution is not expected.

12.3.2 Simple Silicates: Orthosilicates (SiO_4^{4-}) and Pyrosilicates ($Si_2O_7^{6-}$)

The orthosilicate anion is a simple tetrahedron (Figure 12.7). The mineral *olivine* (believed to be a major component of the Earth's mantle), Mg_2SiO_4, contains isolated SiO_4 tetrahedra. The semiprecious stone *garnet*, with composition $(M^{2+})_3(M^{3+})_2(SiO_4^{4-})_3$, also contains isolated SiO_4 tetrahedra.

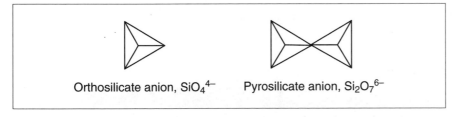

Orthosilicate anion, SiO_4^{4-} Pyrosilicate anion, $Si_2O_7^{6-}$

Figure 12.6 Ortho- and pyro-silicate

Pyrosilicate (similar to pyrophosphate) consists of two tetrahedra with a shared oxygen (Figure 12.6) and is relatively rare.

12.3.3 Ring and Chain Silicates

When SiO_4 tetrahedra share two vertices with each other, rings or chains can be formed, with general composition $[SiO_3^{2-}]_n$. Cyclosilicates are analogues of the cyclic phosphates (Section 12.2), and rings have generally

This type of silicate has been historically called a **metasilicate**.

either three, four or six SiO_4 tetrahedra each with two shared vertices, *e.g.* the **hexametasilicate** anion $[Si_6O_{18}]^{12-}$ (Figure 12.7).

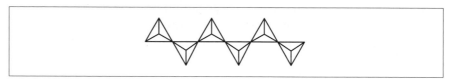

Figure 12.7 The hexametasilicate anion, $[Si_6O_{18}]^{12-}$, which occurs in the mineral beryl, $Be_3Al_2Si_6O_{18}$, the gem form of which is emerald

Silicate chains, such as in Figure 12.8, are found in the **pyroxene** group of minerals; there are many different ways in which the SiO_4 tetrahedra can be arranged in the solid, leading to different minerals. An example is *diopside*, $CaMg(SiO_3)_2$.

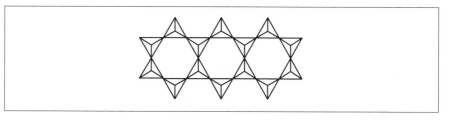

Figure 12.8 A silicate chain in pyroxene minerals: $[SiO_3{}^{2-}]_n$

Amphiboles, unlike pyroxenes, tend to be hydrated minerals, with hydroxyl groups.

Another series of silicates occur which have two pyroxene chains joined together (with SiO_4 tetrahedra sharing two or three vertices, Figure 12.9), giving the **amphibole** minerals, containing $[(Si_4O_{11})^{6-}]_n$ double chains. The *asbestos* range of minerals are examples of amphiboles, for example *tremolite*, $Ca_2Mg_5(Si_4O_{11})_2(OH)_2$.

Figure 12.9 A silicate double chain in amphibole minerals: $[(Si_4O_{11})^{6-}]_n$

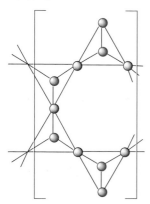

($Si_4O_{11}{}^{6-}$) repeat unit

⬤ Bridging oxygen (Si–O–Si)
   No contribution to charge

⬤ Terminal oxygen (Si–O⁻)
   Contributes 1– charge

Worked Problem 12.5

Q Show that the empirical formula of the silicate framework in amphiboles is $[(Si_4O_{11})^{6-}]_n$.

A In the amphibole structure there are two types of tetrahedra: those sharing two vertices, and those sharing three. Since it is a double chain, the repeat unit must contain two of each type of tetrahedron, with 4 Si atoms and 11 O atoms: Si_4O_{11}. Six of the oxygens are terminal, and the rest are bridging, so the charge is 6–; hence $[(Si_4O_{11})^{6-}]_n$ is correct.

12.3.4 Sheet Silicates

When SiO_4 tetrahedra share three vertices with adjacent tetrahedra, sheet silicates result (Figure 12.10) with the formula $[(Si_4O_{10})^{4-}]_n$. The bonding within these silicate sheets is very strong, but between them is weak, so the compounds tend to cleave into thin sheets. In talc, $Mg_3(OH)_2Si_4O_{10}$, cations only reside between alternate silicate layers, as shown schematically in Figure 12.11. In this structure, there are 'sandwiches' of two $(Si_4O_{10})^{4-}$ sheets with Mg^{2+} and OH^- ions; these silicate–metal–silicate sandwiches then stack together. The bonding between the sandwiches is weak van der Waals forces, so talc is a very soft mineral.

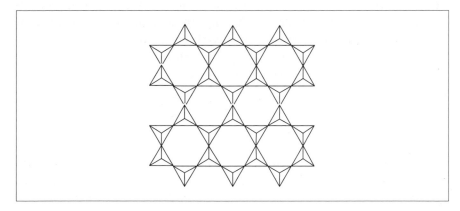

Figure 12.10 A silicate sheet in talc, mica and clay minerals: $[Si_4O_{10}{}^{4-}]_n$

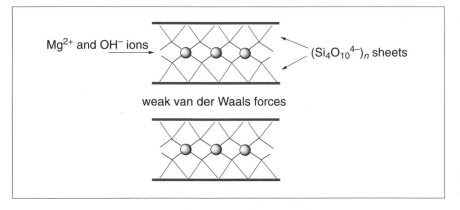

Figure 12.11 The sheet silicate structure of talc

In a related series of important sheet silicate minerals, one in four silicon atoms in the Si_4O_{10} sheet is replaced by an aluminium atom. In order to maintain charge balance an extra 1+ cation is required. These minerals are called micas, such as *phlogopite*, $KMg_3(OH)_2(Si_3AlO_{10})$. In micas, the additional cations (K^+ in *phlogopite*, Li^+ in the lithium ore *lepidolite*) lie between the silicate sheets, and thus they are bonded more strongly by ionic interactions (compared to the weak van der Waals forces of talc); the cleavage of mica is still into thin sheets.

Materials where some silicon atoms are replaced by aluminium are called **aluminosilicates**.

Other important minerals which have sheet silicate structures are the clay minerals.

12.3.5 Network Silicate Materials

If an SiO_4 tetrahedron shares all four of its vertices, a three-dimensional network structure results.

A negative charge is readily introduced into the framework by replacement of some silicon atoms by aluminium atoms (see Section 12.3.4). A commercially important class of network aluminosilicate is the **zeolites**; a large range of both synthetic and natural zeolites is known. One of the basic building blocks for several zeolites is the **sodalite cage** (Figure 12.12), which is typically drawn showing the positions of the Al and Si atoms (rather than the network of linked tetrahedra used for other silicates). By connecting sodalite cages, either through their square faces or their hexagonal faces, regular, large, open cages and channels are generated in a three-dimensional network structure. *Zeolite A* is formed by connecting sodalite cages through square faces (Figure 12.13). Small molecules are able to diffuse into the cavities, but large molecules are excluded. Zeolites are commonly referred to as **molecular sieves** for this reason. Because of the negative charge on the framework (owing to the aluminium-for-silicon substitution), cations are also contained within the lattice. Zeolites can therefore be used as ion-exchange materials; for example, they are commonly added to detergents, where they remove Ca^{2+} (and Mg^{2+}) ions (equation 12.2) which are responsible for hard water formation.

$$Na_2[zeolite]_{(s)} + Ca^{2+}_{(aq)} \rightarrow Ca[zeolite]_{(s)} + 2Na^+_{(aq)} \qquad (12.2)$$

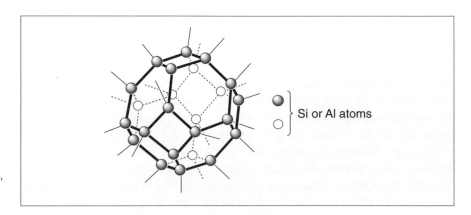

Si or Al atoms

Figure 12.12 The sodalite cage, one of the basic building blocks in a number of zeolites

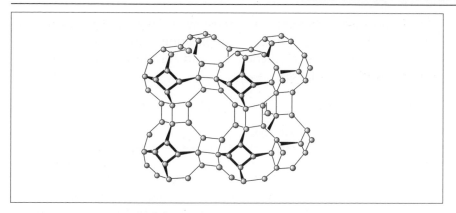

Figure 12.13 The structure of Zeolite A, composed of interconnected sodalite cages

12.4 Silicone Polymers

Organosilicones of composition $(R_2SiO)_n$ are the most important synthetic polymers with a purely inorganic 'backbone'. They are synthesized by hydrolysis of dialkylsilicon dichlorides, *e.g.* Me_2SiCl_2 (Scheme 12.1). The monomeric unit, $Me_2Si=O$, is a highly reactive species, in contrast to the carbon analogue [acetone (ethanone), Me_2CO] so the silicones have polymeric structures with strong Si–O single bonds.

Me_2SiCl_2 is manufactured by the reaction:

$$Si + 2MeCl \xrightarrow[300\ °C]{Cu} Me_2SiCl_2$$

Scheme 12.1 The synthesis of silicone polymers

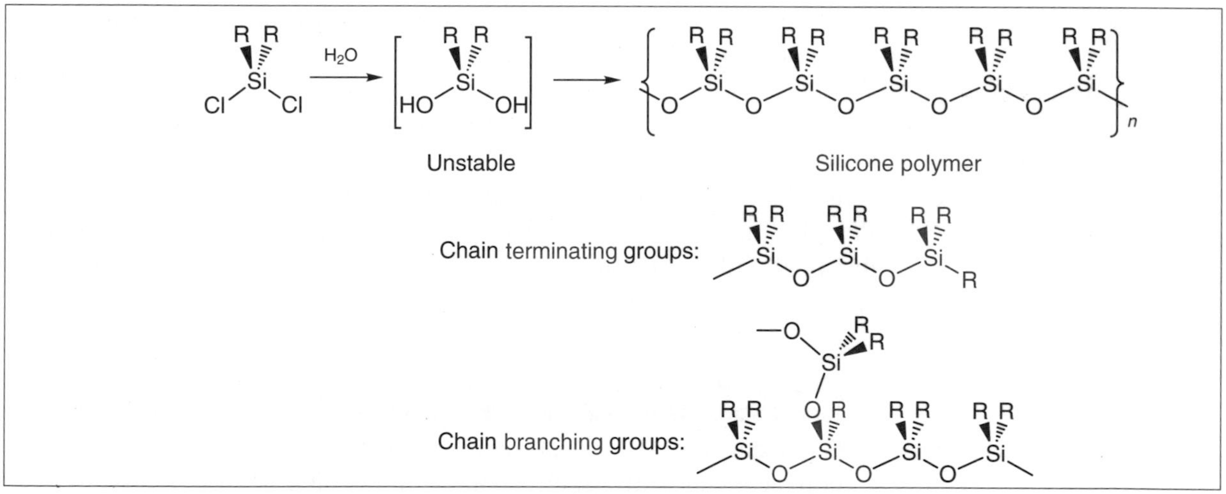

[Scheme 12.1: hydrolysis of R_2SiCl_2 with H_2O to give unstable $R_2Si(OH)_2$ intermediate, then the Silicone polymer. Also shown are Chain terminating groups and Chain branching groups.]

Depending on the conditions used, cyclic siloxanes such as $(Me_2SiO)_3$ (**12.1**) can also be obtained. The chain length of the polymer is moderated by adding some R_3SiCl to the hydrolysis mixture; this forms some terminal $Si–O–SiR_3$ groups (Scheme 12.1), which decreases the viscosity. Similarly, branches in the silicone chains can be incorporated by including some $RSiCl_3$, which increases the viscosity and decreases the solubility. In this way the physical properties of the silicone polymer can be tailored.

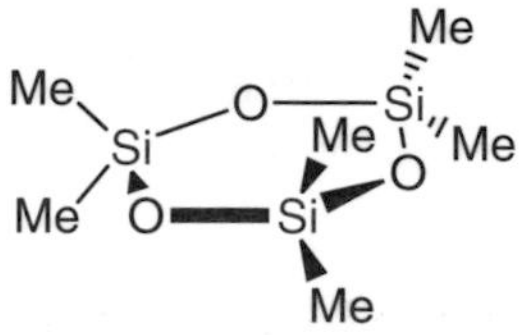

12.1

Silicones are robust materials, resistant to high temperatures and hydrolysis, and find a range of applications, including lubricants, sealants, hydraulic oils, cosmetics, car and furniture polishes, medical implants and contact lenses.

12.5 Polyphosphazenes

12.2

12.3

12.4

12.5

12.6

When PCl_5 and NH_4Cl are heated together in a chlorinated hydrocarbon solvent at around 130 °C, a mixture of cyclic phosphazenes $(Cl_2P=N)_n$ (n = 3 or 4) is formed, and, depending on the conditions, chain compounds of the type $Cl_4P-(N=PCl_2)_n-N=PCl_3$. The cyclic trimer has a planar six-membered ring structure **12.2**, however, the tetramer **12.3** has a puckered eight-membered ring. In both the cyclic trimer and tetramer the P–N bonds are all the same, indicating that delocalized (π) bonding occurs between the filled N p-orbitals and empty acceptor orbitals (which may be d-orbitals) on the P atom.

When the cyclic trimer **12.2** is heated more strongly it polymerizes, forming $(Cl_2P=N)_n$ (**12.4**), a rubbery material. The P–Cl bonds are susceptible to hydrolysis, but can be replaced by nucleophiles such as alkoxides (OR^-) to give **12.5,** or amides (NR_2^-) to give **12.6**, or a wide range of other functional groups, to form more stable polymers with a wide range of properties.

The P=N group is **isoelectronic** with the Si–O group, so the phosphazene materials are isoelectronic with the siloxanes described in Section 12.4. Thus, the trimer $(Cl_2P=N)_3$ (**12.2**) is isoelectronic with $(Me_2SiO)_3$ (**12.1**). Polyphosphazene polymers, like their siloxane counterparts, contain flexible P–N and Si–O bonds, and so maintain their flexible, rubbery properties even at low temperatures.

Worked Problem 12.6

Q Give a combination of main group elements which would form polymers which are isoelectronic with (a) SiO_2; (b) Ge.

A (a) Al has one less electron than Si, and P has one electron more, so $2 \times SiO_2$ ('Si_2O_4') is isoelectronic with $AlPO_4$. Such materials (aluminophosphates) are well known, and like the silicates have complex structures.
(b) Ga has one less electron than Si, and As has one electron more, so GaAs (gallium arsenide) is isoelectronic with Si. Like Si, it is

Summary of Key Points

1. *Many polymeric materials* based on main group elements are known. Examples come from both natural (*e.g.* silicates) and synthetic (*e.g.* polyphosphazene, silicone) systems. The strength of Si–O and P=N bonds accounts for the stability of silicate and silicone, and polyphosphazene materials, respectively.

2. *The SiO_4 tetrahedron* is the basic building block of silicates; these groups can link together by sharing oxygens in many different ways. Replacement of some Si atoms by Al atoms results in the formation of aluminosilicates, which can have different properties to related silicates.

3. *Practical applications* are already widespread for many inorganic-based polymers, and there is an ongoing search to find new materials with interesting and useful properties.

Problems

12.1 Explain what is meant by a *condensation reaction* in the synthesis of a polyphosphate.

12.2 Draw the framework representation for the *tetrametasilicate* anion.

12.3 Show from first principles that the empirical formula of the silicate framework in a pyroxene is $(SiO_3^{2-})_n$.

12.4 Would you expect the following substitutions to occur easily in silicate minerals: (a) Rb^+ (ionic radius 152 pm) for K^+ (138 pm); (b) Ba^{2+} (135 pm) for Ca^{2+} (100 pm).

12.5 What is the main structural difference between the sheet silicate structures of mica and talc, and how is this reflected in their physical properties?

12.6 Draw the structures of the polymeric forms of SO_3 and SeO_2.

12.7 Draw the structure of the polymers formed by: (a) hydrolysis of $Me(Et)SiCl_2$ containing some Et_3SiCl; (b) strongly heating $[N=PMeCl]_3$.

Further Reading

Polymers and the Periodic Table: recent developments in inorganic polymer science, I. Manners, *Angew. Chem., Int. Ed. Engl.*, 1996, **35**, 1602.

Main-group-based rings and polymers, D. P. Gates and I. Manners, *J. Chem. Soc., Dalton Trans.*, 1997, 2525.

Inorganic polymers, N. H. Ray, Academic Press, New York, 1978.

Zeolite molecular sieves, B. M. Lowe, *Educ. Chem.*, 1992, January, 15.

Further Reading

General Inorganic Chemistry

F. A. Cotton, G. Wilkinson, C. A. Murillo and M. Bochmann, *Advanced Inorganic Chemistry*, 6th edn., Wiley-Interscience, New York, 1999.

K. M. Mackay, R. A. Mackay and W. Henderson, *Introduction to Modern Inorganic Chemistry*, 5th edn., Blackie, London, 1996.

D. F. Shriver and P. W. Atkins, *Inorganic Chemistry*, 3rd edn., Oxford University Press, Oxford, 1999.

J. E. Huheey, E. A. Keiter and R. L. Keiter, *Inorganic Chemistry*, 4th edn., HarperCollins, New York, 1993.

C. E. Housecroft and E. C. Constable, *Chemistry – An Integrated Approach*, Longman, Harlow, UK, 1997.

J. D. Lee, *Concise Inorganic Chemistry*, 5th edn., Chapman & Hall, London, 1997.

N. N. Greenwood and A. Earnshaw, *Chemistry of the Elements*, 2nd edn., Butterworth-Heinemann, Oxford, 1997.

D. M. P. Mingos, *Essentials of Inorganic Chemistry*, Oxford University Press, Oxford, 1995.

D. W. Smith, *Inorganic Substances,* Cambridge University Press, Cambridge, 1990.

Multi-volume Reference Series

Encyclopedia of Inorganic Chemistry, ed. R. B. King, Wiley, New York, 1994.

Comprehensive Inorganic Chemistry, ed. J. C. Bailar, H. J. Emeléus, R. S. Nyholm and A. F. Trotman-Dickenson, Pergamon Press, Oxford, 1974.

Comprehensive Coordination Chemistry, ed. G. Wilkinson, R. D. Gillard and J. A. McCleverty, Pergamon Press, Oxford, 1987.

Miscellaneous

G. Aylward and T. Findlay, *SI Chemical Data*, 4th edn., Wiley, Brisbane, 1998.

J. Emsley, *The Elements*, Clarendon Press, Oxford, 1989.

J. D. Woollins, *Inorganic Experiments*, VCH, Weinheim, 1995.

Main Group Chemistry

A. G. Massey, *Main Group Chemistry*, Ellis Horwood, Chichester, 1990.

C. E. Housecroft, *Cluster Molecules of the p-Block Elements*, Oxford University Press, Oxford, 1994.

J. D. Woollins, *Non-metal Rings, Cages and Clusters*, Wiley, Chichester, 1988.

Solid-state Chemistry (includes silicate materials)

U. Müller, *Inorganic Structural Chemistry,* Wiley, Chichester, 1993.

A. F. Wells, *Structural Inorganic Chemistry*, 5th edn., Oxford University Press, Oxford, 1984.

D. M. Adams, *Inorganic Solids*, Wiley, New York, 1974.

M. T. Weller, *Inorganic Materials Chemistry*, Oxford University Press, Oxford, 1994.

Organometallic Chemistry

C. Elschenbroich and A. Salzer, *Organometallics*, VCH, Weinheim, 1989.

Comprehensive Organometallic Chemistry, ed. G. Wilkinson, F. G. A. Stone and E. W. Abel, Pergamon, Oxford, 1982; *Comprehensive Organometallic Chemistry II*, ed. G. Wilkinson, F. G. A. Stone and E. W. Abel, Elsevier, Oxford, 1995.

Periodicity and the Periodic Table

D. M. P. Mingos, *Essential Trends in Inorganic Chemistry*, Oxford University Press, Oxford, 1998.

N. C. Norman, *Periodicity and the s- and p-Block Elements*, Oxford University Press, Oxford, 1997.

Industrial Chemistry

T. W. Swaddle, *Inorganic Chemistry: an Industrial and Environmental Perspective*, Academic Press, San Diego, 1997.

W. Büchner, R. Schliebs, G. Winter and K. H. Büchel, *Industrial Inorganic Chemistry*, VCH, Weinheim, 1989.

Answers to Problems

Chapter 1

1.1. (a) Be; (b) N (see Section 1.5); (c) N (see Section 1.5); (d) Se$^+$ (it is much harder to remove an electron from a positive ion than a neutral atom); (e) K (ionization energies decrease going down a group).

1.2. (a) S has six valence electrons, plus six from σ-bonds to F = a total of 12 electrons, or six pairs; therefore shape is a regular octahedron.
(b) Se has six valence electrons, plus two from σ-bonds to F = a total of eight electrons, or four pairs. Shape is angular, two bonds and two lone pairs. Expect F–Se–F bond angle < tetrahedral angle of 109.5° because of lone pairs on Se.
(c) C is central atom; structure is:

$$\text{HO}-\overset{\displaystyle O^-}{\underset{\displaystyle O}{C}}$$

C has four valence electrons, plus three electrons from σ-bonds to three C atoms. Negative charge on O so ignore, but subtract one electron for C=O π-bond. Total = six electrons, or three pairs; therefore trigonal. Consider resonance hybrid:

$$\left[\ \text{HO}-\overset{\displaystyle O}{\underset{\displaystyle O}{C}}\ \right]^-$$

Therefore the O–C–O bond angle will be slightly less than 120°, and the HO–C–O bond angles slightly greater.

(d) Xe has eight valence electrons, σ-bonds to O and four F contribute five electrons. Subtract one electron for Xe=O π-bond. Total = 12 electrons or six pairs; structure based on octahedron with one lone pair. Lone pair and double bond occupy more space in valence shell than fluorines, so expect these groups to be opposite each other, with F–Xe–F and O–Xe–F bond angles approximately 90°.
(e) P is central atom, with five valence electrons. F and Cl atoms contribute a total of five electrons, so total = 10, or five pairs. Structure based on trigonal bipyramid. Expect Cl atoms to prefer equatorial and F atoms to prefer axial positions, so the structure is:

(f) S has six valence electrons, and σ-bonds to F and Cl contribute three electrons, but subtract one for positive charge. Total = eight electrons, or four pairs; therefore based on tetrahedron. Shape is trigonal pyramid with lone pair on S.
(g) Structure is:

Consider one S as a central atom. S has six valence electrons, σ-bonds to two O and one S contribute three electrons. Negative charge on O so ignore, but subtract one electron for S=O π-bond. Total is eight electrons or four pairs; therefore trigonal pyramid shape at S with lone pair. All O atoms identical through resonance, so expect partial double bond character in S–O bonds, and O–S–O bond angle >109.5°.

1.3. BF_3 has three electron pairs, and is trigonal planar with F–B–F bond angles of 120°. The other molecules all have four electron pairs in the central atom valence shell, and are either regular tetrahedral (CF_4), or have bond angles smaller than 109.5° (SF_2, PF_3, H_2S) owing to lone pair–bonding pair repulsion.

1.4. In O_2^{2+} the π_2^* MO will be empty, and so the bond order will be 3. The species is isoelectronic with N_2.

1.5. (a) Antibonding. (b) Non-bonding.

Chapter 2

2.1. Hydrogen in combination with electronegative elements has the +1 oxidation state, *e.g.* HCl, H_2O, NH_3. In combination with electropositive elements, hydrogen has the –1 oxidation state, *e.g.* NaH, CaH_2.

2.2. (a) I; (b) C; (c) B; (d) O.

2.3. Recall from Sections 2.4.1 and 2.4.2 that, for compounds containing E–H bonds, if electronegativity of E is <1.2 then ionic, from 4.0 to 1.5 then covalent.
(a) CsH is formed from the very electropositive element caesium, so it will be ionic, Cs^+H^-.
(b) P is a non-metal (electronegativity 2.2), so PH_3 is covalent.
(c) B is likewise a non-metal (electronegativity 2.0), so B_2H_6 is covalent.
(d) $NaBH_4$ is an ionic solid, $Na^+[BH_4]^-$, but the B–H bonds are covalent.
(e) Cl is an electronegative non-metal (electronegativity 3.16), so HCl will be covalent, but the H–Cl bond will be polarized $H^{\delta+}$–$Cl^{\delta-}$.

2.4. Although H_2S_2 has a higher molecular mass than H_2O_2, in the latter there will be extensive H⋯O⋯H hydrogen bonding. These intermolecular forces must be overcome for the liquid to boil, so H_2O_2 will have a higher boiling point than H_2S_2. The situation is the same as for H_2O (b.p. 100 °C) and H_2S (b.p. –60 °C).

Chapter 3

3.1. (a) Zero (by definition); (b) +1 (contains the PF_6^- anion); (c) +1 (contains the O_2^- anion); (d) this compound is a salt, [Na(18-crown-6)]$^+$Na$^-$, so the oxidation states are +1 and –1 respectively.

3.2. (a) Li; (b) K, Rb or Cs; (c) Na; (d) Li.

3.3. (a) Going down the group the size of the atom increases, as electrons fill successive shells (electron energy levels). For Cs, the valence 6s electron is further from the nucleus, is 'shielded' by completed 'shells', and is more easily ionized.
(b) Similarly, the Cs^+ ion is much larger than the Li^+ ion, its charge density is much lower, and the ion–dipole interactions with water

molecules are much weaker. Cs^+ therefore has a lower hydration energy than Li^+.

3.4. (a) Dissolve Li metal in water, giving LiOH:

$$2Li + 2H_2O \rightarrow 2LiOH + H_2$$

then pass CO_2 through the LiOH solution:

$$2LiOH + CO_2 \rightarrow Li_2CO_3 + H_2O$$

Li_2CO_3 is sparingly soluble in water (1.3 g in 100 cm^3 water), so will precipitate out.
(b) React bromobenzene (PhBr) with Li metal in dry diethyl ether solvent:

$$PhBr + 2Li \rightarrow PhLi + LiBr$$

(c) Dissolve Li metal in anhydrous liquid ammonia, giving Li^+ e^-(solv), which is converted to $LiNH_2$ by addition of a transition metal compound catalyst, *i.e.*:

$$2Li + 2NH_3 \rightarrow 2LiNH_2 + H_2$$

Alternatively, pass anhydrous gaseous NH_3 through a solution of PhLi (NH_3 is a stronger acid than benzene):
$PhLi + NH_3 \rightarrow PhH + LiNH_2$

3.5. Statement (d) is incorrect: potassium has a maximum oxidation state of +1, and K_2O_2 is formulated as $(K^+)_2(O_2^{2-})$.

3.6. Since all salts are of the type M^+Br^-, lattice energy decreases with increasing ionic radii (Box 3.1), so CsBr is expected to have the lowest lattice energy, and hence the lowest melting point.

3.7. Li_2CO_3, $LiNO_3$, Li_2O_2.

3.8. (a) $r^+/r^- = 149/220 = 0.68$, predicts NaCl structure; (b) $r^+/r^- = 102/220 = 0.46$, predicts NaCl structure; (c) $r+/r^- = 74/218 = 0.34$, predicts sphalerite (ZnS) structure.

Chapter 4

4.1. (a) Zero, by definition; (b) +2 (contains C_2^{2-}, ethynide anion); (c) +2 (H_2O ligands are neutral).

4.2. (a) Ba; (b) Be; (c) Mg.

4.3. The decomposition of a salt is promoted by the formation of a new compound with a high lattice energy. For $BeCO_3$ and $BeSO_4$, BeO will be formed by decomposition. The small doubly charged ions will result in a very high lattice energy for BeO which will help to drive the reaction (the reaction is also favoured by an increase in entropy, since a gaseous product is formed).

4.4. In the formation of a nitride from the metal and elemental nitrogen (N_2), the two most important thermochemical factors which must be considered are the strong $N\equiv N$ bond which must be broken, and the lattice energy of the resulting nitride salt. Lattice energy (see Box 3.1) is maximum for highly charged and small ions. The Group 2 nitrides all exist because the lattice energies of the M_3N_2 compounds are very large. For lithium, the small size of the Li^+ ion stabilizes Li_3N, but for the other (larger) Group 1 cations the lattice energy is smaller, and the compounds M_3N are less stable. In fact, calculations suggest Na_3N should be stable but it has never been isolated.

4.5. (a) $BaO_{2(s)} + H_2SO_{4(aq)} \rightarrow BaSO_{4(s)} + H_2O_{2(l)}$
(b) $Ba(NO_3)_{2(aq)} + Na_2SO_{4(aq)} \rightarrow BaSO_{4(s)} + Na_2SO_{4(aq)}$
(c) $Ca_{(s)} + H_{2(g)} + heat \rightarrow CaH_{2(s)}$

4.6. (a) Add a Grignard reagent to a C=O bond, then acidify. Several possibilities, *e.g.*
Me–C(O)–Et + PhMgBr → Me–C(OMgBr)(Et)(Ph) → MeC(Et)(OH)Ph
or Me–C(O)–Ph + EtMgBr → *etc.*
(b) $AsCl_3 + 3PhMgBr \rightarrow AsPh_3 + 3MgBrCl$

Chapter 5

5.1. B_4Cl_4, AlCl, TlCl

5.2. (a) B; (b) Al; (c) B (BF_5^{2-} cannot form); (d) Ga; (e) Tl.

5.3. (a) See Section 5.1; (b) see Section 4.3.

5.4. (a) $BBr_3 + 3H_2O \rightarrow B(OH)_3 + 3HBr$ (hydrolysis reaction)
(b) $BCl_3 + Me_4N^+Cl^- \rightarrow Me_4N^+BCl_4^-$ (adduct formation, Lewis acid = BCl_3, base = Cl^-)

(c) $Ph_2PCl + Li[AlH_4] \rightarrow Ph_2PH + Li[AlH_3Cl]$ (reduction reaction). Alternatively, $4Ph_2PCl + Li[AlH_4] \rightarrow 4Ph_2PH + Li[AlCl_4]$

5.5. The fluorides of many elements tend to have high melting points because they have a significant degree of ionic character compared to the other halides of the same element. AlF_3 is a *polymeric solid* built up from AlF_6 octahedra, containing Al–F–Al bridges. In contrast, aluminium bromide is a *molecular substance*, Al_2Br_6. The attraction between molecules is relatively weak, so the compound has a low melting point and is soluble in the low polarity solvent benzene.

5.6. (a) Both BCl_3 and B_2H_6 are formally electron deficient (less than the octet of electrons); (b) $H_3N \rightarrow BF_3$.

5.7. (a) $[B_6H_6]^{2-}$. Total electron count $= (6 \times 3) + (6 \times 1) + 2(\text{charge}) = 26$; subtract (6×2) electrons for BH units, gives 14 electrons, or 7 SEP; the structure is based on an octahedron; since there are 6 B atoms, the structure is *closo*: see Figure 5.11a.
(b) $B_{10}C_2H_{12}$. Total electron count $= (10 \times 3)(B) + (2 \times 4)(C) + (12 \times 1)(H) = 50$ electrons; subtract (12×2) electrons for 4 BH and 2 CH units $= 26$ electrons, or 13 SEP; the structure is based on a 12-vertex polyhedron: the *icosahedron* (Figure 5.2). Since there are 12 B and C atoms, the structure is *closo*. There are three possible isomers of the product; the *para* form, with the two carbons as far apart as possible, is actually the most stable:

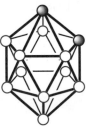 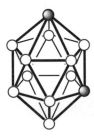 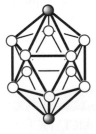

ortho-carborane   *meta*-carborane   *para*-carborane

○ ○ } B
● C

ortho, *meta* and *para* isomers of carborane, $B_{10}C_2H_{12}$. H atoms are omitted for clarity

(c) B_4H_{10}. Total electron count $= (4 \times 3) + (10 \times 1) = 22$; subtract (4×2) electrons for BH units gives 14 electrons, or 7 SEP; hence the structure is derived from an octahedron; there are only 4 B

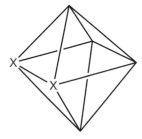

An octahedron, showing the positions of the two missing vertices (X) in B_4H_{10}

atoms, so 2 missing vertices, therefore an *arachno* cluster. Wade's rules give no indication as to *which* vertices are missing, and the actual structure is that with adjacent missing vertices.

5.8. In solution the Al^{3+} is solvated: $[Al(H_2O)_6]^{3+}$. Owing to the high charge density of the ion, hydrolysis occurs, to reduce the charge, by conversion of a coordinated H_2O ligand into a coordinated OH^- ion and free H^+ (which makes the solution acidic). The Tl^+ ion has a much lower charge density, is less strongly solvated, and there is no driving force for the hydrolysis reaction to proceed.

See equation 5.21

5.9. $B_2O_3 + 3Mg \rightarrow 2B + 3MgO$

5.10. Compare Worked Problem 5.5.
(a) $\Delta H = (-598) + (-1273) + 4(-286) - 2(-189) = -2637$ kJ mol^{-1}. 2 mol of $LiBH_4$ has a mass of 43.6 g, so the energy available is -60.5 kJ g^{-1}.
(b) $2B_4H_{10} + 11O_2 \rightarrow 4B_2O_3 + 10H_2O$
$\Delta H = 4(-1273) + 10(-286) - 2(+66) = -7828$ kJ mol^{-1}. 2 mol of B_4H_{10} has a mass of 53.2g, so the energy available is -147 kJ g^{-1}.
(c) $B_{10}H_{14} + 11O_2 \rightarrow 5B_2O_3 + 7H_2O$
$\Delta H = 5(-1273) + 7(-286) - (+32) = -8335$ kJ mol^{-1}. 1 mol of $B_{10}H_{14}$ has a mass of 122.1 g, so the energy available is -68 kJ g^{-1}.
Thus, combustion of 1 g of B_4H_{10} will produce the most energy. Compared to hydrogen ($\Delta_f H°$ of water is -286 kJ mol^{-1} for combustion of 1 mole, or 2 g of H_2) which produces -143 kJ g^{-1}, B_4H_{10} is a very useful source of energy.

Chapter 6

6.1. (a) +4; (b) +2; (c) +4; (d) +3 (the Pb–Pb bond does not contribute to the oxidation state).

6.2. (a) Si; (b) C; (c) Pb; (d) C.

6.3. Carbon oxides are gaseous, with C=O $p\pi$–$p\pi$ bonding; silicon oxides and oxyanions are predominantly polymeric, with strong Si–O single bonds. Carbon halides are stable to hydrolysis; silicon halides hydrolyse rapidly and act as Lewis acids, forming adducts with Lewis bases, with coordination number 5 or 6 (carbon is restricted to 4). Long-chain carbon compounds are common and stable (organic compounds), often with C=C and C≡C bonds; long-chain silicon compounds are less common, and Si=Si and Si≡Si bonds are rare.

6.4. Going down the group the E atom becomes much larger, so the E–H bonds become progressively weaker and the compounds are less stable with respect to decomposition to the elements.

6.5. (a) See Section 6.1; (b) Section 6.2.1; (c) Section 6.2.1; (d) Section 6.5.2.

6.6. (a) I_2 oxidizes the tin to the IV oxidation state, giving SnI_4: $Sn + 2I_2 \rightarrow SnI_4$.
(b) $Be_2C + 4H_2O \rightarrow CH_4 + 2Be(OH)_2$. A hydrolysis reaction.
(c) $CCl_4 + H_2O \rightarrow$ No reaction. The C–Cl bonds of CCl_4 are kinetically stable towards hydrolysis.
(d) $Et_2SiCl_2 + Li[AlH_4] \rightarrow Et_2SiH_2 + Li[AlH_2Cl_2]$. A reduction reaction to give the hydride.

6.7. As with the germane in Worked Problem 6.3, different branched-chain isomers can occur.

Hydrogen atoms not shown

Chapter 7

7.1. (a) –2; (b) +5; (c) +5; (d) +1; (e) phosphorus pentoxide (P_2O_5) is actually P_4O_{10} (definitely not PO_5!), so the oxidation state is +5.

7.2. (a) N; (b) P; (c) N; (d) Bi.

7.3. (a) $NH_4NO_{2(s)} \rightarrow N_{2(g)} + 2H_2O_{(g)}$
(b) $NH_4NO_{3(s)} \rightarrow N_2O_{(g)} + 2H_2O_{(g)}$
(c) $Zn_3As_{2(s)} + 6HCl_{(aq)} \rightarrow 3ZnCl_{2(aq)} + 2AsH_{3(g)}$
(d) $As_2O_3 + 6Zn + 12H^+ \rightarrow 2AsH_3 + 6Zn^{2+} + 3H_2O$
(e) $P_4O_{10} + 6H_2O \rightarrow 4H_3PO_4$

7.4. The boiling point of NH_3 is higher than that expected by simply extrapolating the boiling points of the other hydrides, owing to strong N–H⋯N hydrogen bonding.

7.5. Using VSEPR, the molecules have 3 electron pairs on the central N atom, a double bond (to O), a lone pair and a bonding pair to X. Because of the higher electron density of the double bond and closer proximity of the lone pair to the nitrogen nucleus, the repulsion between them is greater, so the X–N–O angle should be decreased from the regular angle of 120° expected between three bonding pairs. However, fluorine is highly electronegative, so the space occupied near the N by the N–F bonding pair is decreased, and the F–N–O angle is further decreased.

7.6. For P≡P to be stable relative to three P–P single bonds, the bond energy of the P≡P bond should be at least three times that of the P–P single bond; this is clearly not the case. Similarly, the P=P bond energy is less than double the P–P single bond energy. Thus, P=P and P≡P triple bonds are unstable relative to P–P single bonds.

Chapter 8

8.1. (a) –2; (b) +2; (c) +6; (d) +5; (e) –1.

8.2. (a) O; (b) Te; (c) Se; (d) O; (e) Po.

8.3. The element itself (Section 8.2.2), sulfanes HS_nH (Section 8.3.5), chlorosulfanes ClS_nCl (Section 8.4.2), polysulfides S_n^{2-} (Section 8.3.5), polythionic acids $[O_3S-S_n-SO_3]^{4-}$ (Section 8.7.1).

8.4. (a) Heat the hydrated $ZnBr_2$ in excess sulfur dibromide oxide (thionyl bromide), $SOBr_2$:
$ZnBr_2.2H_2O + 2SOBr_2 \rightarrow ZnBr_2 + 2SO_2 + 4HBr$
(b) React benzoic acid with hydrogen peroxide:
$PhC(O)OH + H_2O_2 \rightarrow PhC(O)OOH + H_2O$
(c) React sulfur with sodium metal; this is best achieved by dissolving the sodium in liquid ammonia (see Section 3.6), then adding the stoichiometric amount of sulfur:
$2Na + 3S \rightarrow Na_2S_3$
Then $Na_2S_3 + 2HCl \rightarrow H_2S_3 + 2NaCl$

8.5. AlS_2 (a) since this will either be a compound of aluminium(IV) (with two S^{2-} ions) or of aluminium(II) (with an S_2^{2-} ion). Al(III) is the only stable oxidation state. CaS and BaS all contain the S^{2-} ion, CS_2 is a stable molecular substance (analogous to carbon dioxide, CO_2) and K_2S_2 contains the S_2^{2-} ion (which is also found in minerals such as pyrite, FeS_2, historically known as 'fool's gold' because of its metallic appearance.

8.6. (a) A precipitate of black Ag_2S will form:
$$2AgNO_{3(aq)} + H_2S_{(g)} \rightarrow Ag_2S_{(s)} + 2HNO_{3(aq)}$$
(b) A redox reaction will occur:
$$2S_2O_3^{2-} + Br_2 \rightarrow S_4O_6^{2-} + 2Br^-,$$ so the orange colour of bromine will disappear.

8.7. (a) $2H_2S + 3O_2 \rightarrow 2SO_2 + 2H_2O$
(b) $H_2S + 4O_2F_2 \rightarrow SF_6 + 2HF + 4O_2$

8.8. (a) $2H_2S \rightarrow 2S + 4H^+ + 4e^-$ $E°$ –0.14 V
$SO_2 + 4H^+ + 4e^- \rightarrow S + 2H_2O$ $E°$ +0.45 V
Adding: $2H_2S + SO_2 + 4H^+ \rightarrow 3S + 4H^+ + 2H_2O$ E +0.31 V
i.e. $2H_2S + SO_2 \rightarrow 3S + 2H_2O$. The value of E is positive so the reaction will proceed.
(b) $S_2O_3^{2-} + 6H^+ + 4e^- \rightarrow 2S + 3H_2O$ $E°$ +0.5 V
$S_2O_3^{2-} + H_2O \rightarrow 2SO_2 + 2H^+ + 4e^-$ E –0.4 V (note sign)
Adding: $2S_2O_3^{2-} + 6H^+ + H_2O \rightarrow 2S + 2SO_2 + 3H_2O + 2H^+$ E +0.1 V
i.e. $S_2O_3^{2-} + 2H^+ \rightarrow SO_2 + S + H_2O$ E +0.1 V
The value of E for the reaction is positive, so the reaction will spontaneously proceed under standard conditions, although concentration effects can alter $E°$ values. Thiosulfate is unstable in acid solution.

Chapter 9

9.1. (a) +7; (b) Cl +3, F –1; (c) +2; (d) +5; (e) +1.

9.2. (a) Cl; (b) I; (c) F; (d) Br.

9.3. The first ionization energy decreases going down the group because the valence electron to be ionized is further from the nucleus: the atomic radius increases. The enthalpy of vaporization increases because the heavier molecules have increased intermolecular forces. The bond energy of the X_2 molecule decreases because of the poorer overlap between large diffuse p-orbitals in iodine compared to chlorine.

9.4. (a) $Cl_2 + IO_3^- + 2OH^- \rightarrow IO_4^- + 2Cl^- + H_2O$
(b) $2KMnO_4 + 10KCl + 8H_2SO_4 \rightarrow 2MnSO_4 + 6K_2SO_4 + 8H_2O + 5Cl_2$

9.5. BrF_5 is able to donate a fluoride ion to AsF_5 (a powerful fluoride acceptor), giving the ions BrF_4^+ and AsF_6^-, which are able to conduct.

9.6. BCl_3, $SiCl_4$ and PCl_5 are all rapidly hydrolysed, because there is a low-energy pathway to hydrolysis involving coordination of water to the central atom, while CCl_4 and SF_6 are stable to hydrolysis because the C and S atoms are sterically protected from attack by an incoming water molecule. The C atom has no available orbitals for an incoming water molecule to interact with.

9.7. (a) $CsF + ClF_3 \rightarrow Cs^+[ClF_4]^-$
(b) $CsF + BrF_5 \rightarrow Cs^+[BrF_6]^-$

9.8. Other examples of molecular compounds which ionize in the solid state are PCl_5 ($PCl_4^+PCl_6^-$), SCl_4 ($SCl_3^+Cl^-$) and N_2O_5 ($NO_2^+NO_3^-$).

9.9. Elemental astatine would be reduced to the At^- ion using a reducing agent, and mixed with iodide ions, I^-. If Ag^+ ions are added, a precipitate of AgI will form, which is insoluble in concentrated aqueous NH_3. If astatine was behaving the same as iodine, the AgI precipitate will contain the radioactive At.

9.10.

Lewis dot diagram for I_3^-

Consider the central I atom, with 7 valence electrons. Add two electrons for the two σ-bonds, plus one for the negative charge, giving 10 electrons, or 5 pairs. Hence the shape is derived from a trigonal bipyramid with equatorial lone pairs. The two I atoms bonded to the central I are in axial positions, so the ion is linear (see Figure 9.8).

Recall from Section 1.4.5 that lone pairs in trigonal bipyramids prefer equatorial positions.

9.11. Compare with the chemistry of the related chlorine species under the same conditions:

(a) See equation 9.16; hence $(CN)_2 + 2OH^- \rightarrow CN^- + CNO^- + H_2O$

(b) A precipitate of sparingly soluble AgCN would be formed, analogous to the reaction $Ag^+_{(aq)} + Cl^-_{(aq)} \rightarrow AgCl_{(s)}$

(c) A mixture of $I_2 + Cl_2$ gives the interhalogen ICl (Section 9.7.1), so the reaction of $I_2 + (CN)_2$ is expected to give I–CN.

(d) Since the reaction of $Cl_2 + H_2$ gives HCl, the reaction $H_2 + (CN)_2 \rightarrow 2HCN$ is expected.

(e) Cl_2 is a strong oxidizing agent, forming Cl^- ions; hence $(CN)_2$ should behave similarly, giving CN^- ions.

9.12. In basic solution:

$Cl_2 + 2e^- \rightarrow 2Cl^-$ $E°$ +1.36 V

$2ClO^- + 2H_2O + 2e^- \rightarrow Cl_2 + 4OH^-$ $E°$ +0.42 V

Hence:

$Cl_2 + 2e^- \rightarrow 2Cl^-$ $E°$ +1.36 V

$Cl_2 + 4OH^- \rightarrow 2ClO^- + 2H_2O + 2e^-$ E –0.42 V (note sign)

Adding:

$Cl_2 + 2OH^- \rightarrow Cl^- + ClO^- + H_2O$ E +0.94 V

The reaction is therefore spontaneous in basic solution, under standard conditions.

In acid solution:

$Cl_2 + 2e^- \rightarrow 2Cl^-$ $E°$ +1.36 V

$2HClO + 2H^+ + 2e^- \rightarrow Cl_2 + 2H_2O$ $E°$ +1.63 V

Thus for the reaction

$Cl_2 \rightarrow Cl^- + HClO + H^+$ $E°$ –0.27 V

so the reaction does not proceed spontaneously.

Chapter 10

10.1. (a) +6; (b) +6; (c) +2; (d) +6.

10.2. (a) Ne; (b) Ar; (c) Kr; (d) Rn.

10.3. (a) $C_6F_5I + XeF_2 \rightarrow C_6F_5IF_2 + Xe$

(b) $Ph_2S + XeF_2 \rightarrow Ph_2SF_2 + Xe$

(c) $S_8 + 24XeF_2 \rightarrow 8SF_6 + 24Xe$

10.4. (a) $KrF_2 + B(OTeF_5)_3 \rightarrow Kr(OTeF_5)_2 +$ other products.

(b) Either addition of F^- to $XeOF_4$: $XeOF_4 + F^- \rightarrow [XeOF_5]^-$ or hydrolysis of $[XeF_7]^-$: $[XeF_7]^- + H_2O \rightarrow [XeOF_5]^- + HF$

For 10.4 (a), see J. C. P. Sanders and G. J. Schrobilgen, *J. Chem. Soc., Chem. Commun.*, 1989, 1576.

For 10.4 (b), compare the hydrolysis of XeF_6, Section 10.5; see A. Ellern and K. Seppelt, *Angew. Chem., Int. Ed. Engl.*, 1995, 34, 1586.

10.5. $Xe + O_2F_2 \rightarrow XeF_4 + 2O_2$

10.6. Using the method as in Worked Problem 10.1:
(a) Cs^+Xe^-: $\Delta_f H° = 79 + 378 + 43 - 457 = +43$ kJ mol^{-1}.
(b) K^+Ne^-: $\Delta_f H° = 90 + 421 + 31 - 613 = -71$ kJ mol^{-1}.
Thus, potassium neonide might be stable, but caesium xenonide (with a positive enthalpy of formation) is probably unstable.

10.7. (a) Xe has 8 valence electrons, plus 5 electrons from σ-bonds to F, plus 1 electron (for the negative charge), giving a total of 14 electrons, or 7 pairs. There are several geometries that can be adopted by seven-electron-pair species (see Figure 1.6). $[XeF_5]^-$ adopts the pentagonal bipyramidal arragement, since it allows the two lone pairs to adopt positions as far apart as possible:

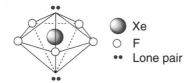

Xe
F
Lone pair

Structure of $[XeF_5]^-$

See K. O. Christe, E. C. Curtis, D. A. Dixon, H. P. Mercier, J. C. P. Sanders and G. J. Schrobilgen, *J. Am. Chem. Soc.*, 1991, 113, 3351.

(b) Cl has 7 valence electrons, plus two for σ-bonds to Xe, minus one electron for the positive charge, giving 8 electrons, *i.e.* 4 pairs. The Cl has 2 bonding pairs and two lone pairs, and is therefore bent, with an experimentally determined Xe–Cl–Xe angle of about 117°.

Chapter 11

11.1. (a) +2; (b) +2; (c) +1; (d) +2.

11.2. (a) Hg; (b) Hg; (c) Zn.

11.3. (a) $Zn_{(s)} + 2HCl_{(aq)} \rightarrow ZnCl_{2(aq)} + H_{2(g)}$
(b) $Zn_{(s)} + 2NaOH_{(aq)} + 2H_2O \rightarrow Na_2[Zn(OH)_4]_{(aq)} + H_{2(g)}$
(c) $Zn_{(s)} + 2NH_4Cl_{(NH_3)} \rightarrow ZnCl_{2(NH_3)} + H_{2(g)} + 2NH_{3(l)}$
(d) $Zn_{(s)} + 2NaNH_{2(NH_3)} + 2NH_{3(l)} \rightarrow Na_2[Zn(NH_2)_4]_{(NH_3)} + H_{2(g)}$

11.4. The metallic elements in the bottom right-hand side of the p-block: As, Sb, Bi, Sn, Pb.

11.5. (a) Since HgS is a very stable, insoluble solid, the Hg_2^{2+} ion will disproportionate:

$$Hg_2^{2+} + S^{2-} \rightarrow HgS + Hg$$

(b) Since carbonates of 2+ metal ions are insoluble, $CdCO_3$ will precipitate:

$$Cd^{2+} + CO_3^{2-} \rightarrow CdCO_{3(s)}$$

Chapter 12

12.1. The term condensation reaction refers to the elimination of a small molecule (H_2O, HCl, *etc.*) from two precursor species. In polyphosphate formation, phosphoric acid or hydrogenphosphate anions (HPO_4^{2-} or $H_2PO_4^-$) are heated and water is eliminated, forming a P–O–P linkage.

12.2. The tetrametasilicate anion contains four Si atoms ('tetra') and is cyclic. The framework representation is the same as that of the tetrametaphosphate anion $[P_4O_{12}]^{4-}$ (Figure 12.2).

12.3. In the pyroxene structure the repeat unit is based on an SiO_3 group. This has two terminal oxygens (each of which bear a negative charge) and one bridging oxygen (which bears no charge), so the $(SiO_3^{2-})_n$ formula is correct.

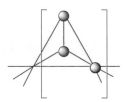

(SiO_3^{2-}) repeat unit

◐ Bridging oxygen (Si–O–Si)
No contribution to charge

◓ Terminal oxygen (Si–O⁻)
Contributes 1– charge

12.4. (a) These ions are just within the 10% 'limit' in the difference in ionic radii, so they would be expected to substitute each other.
(b) No, the Ba^{2+} ion is too large to effectively substitute for Ca^{2+}.

12.5. In talc, the sheets are composed only of silicon and oxygen, and there are sandwiches of two such sheets with Mg^{2+} cations. The bonding between the sandwiches is weak, so talc is a soft mineral. In mica, some of the Si atoms are replaced by Al atoms plus monovalent cations. The $(Al^{3+} + M^+)$ for Si^{4+} substitution means that the aluminosilicate layers are negatively charged, and the M^+ cations reside between the sandwiches. These bond the sandwiches together more tightly, by ionic interactions, so mica is a harder mineral than talc, but still cleaves in thin sheets.

12.6. Refer to Sections 8.5.1 and 8.5.2.

12.7. (a) This will give a poly(siloxane) based on the [Me(Et)Si–O–] repeat unit. The Et_3SiCl will form a chain-terminating group. The structure of the polymer will be:

(b) This will give a polyphosphazene containing the [MeClP=N] repeat unit, *i.e.*

Subject Index